SEASONS
STORIES OF FAMILY, GOD
AND THE
GREAT OUTDOORS

ENDORSEMENTS

A lifetime spent in the woods and on the water, even if not every day, can evoke thoughts, emotions and even a spirituality not easily attained by others. Larry has done a wonderful job of relating all of these in this wonderful book. He brings a peaceful understanding of why generation after generation has found answers to life's hardest questions by absorbing the sights, sounds and feelings experienced while surrounded by nature. Here is to a life well lived. Cheers, Larry!
– Mark C. Smith, Executive Director, Association of Great Lakes Outdoor Writers

"I met Larry more than 30 years ago. Soon, he became one of my favorite writers. His writing will touch your heart, will make you laugh, and might even make you cry. But mostly, he will make you glad you are alive.
– Bobby Whitehead, Editor Emeritus Outdoor Guide Magazine

Larry tugs at every heart-warming emotion when he details a beautiful morning sunrise from a deer stand, the smells, sounds and visual impact of a crackling campfire and the importance of skipping rocks with grandkids. I recommend reading this book one story a night, to enjoy and savor every outdoor memory and moment he shares.
– Tim Huffman, *award-winning outdoor writer, author, photographer.*

SEASONS
STORIES OF FAMILY, GOD
and the
GREAT OUTDOORS

LARRY WHITELEY

Copyright © 2021 Larry Whiteley
All cover art copyright © 2021 Larry Whiteley
All Rights Reserved

No part of this book may be reproduced or transmitted in any form or by any means, electronic or mechanical, including photocopying, recording, or by any information storage and retrieval system, without permission in writing from the author.

A portion of the proceeds from this publication will be donated to organizations that teach kids and families to enjoy the great outdoors.

Publishing Coordinator – Sharon Kizziah-Holmes

Paperback-Press
an imprint of A & S Publishing
A & S Holmes, Inc.

ISBN -13: 978-1-951772-75-8

TABLE OF CONTENTS

Endorsements ii
Dedication ix
Acknowledgments x
Introduction i
SPRING ii
    The Eagle Sees the Round Rainbow 1
    Signs of Spring 5
    Grandpa's Old Shotgun 9
    The Sounds of Nature 12
    Daydreams and Night Dreams 16
    A Text from a Grandson 20
    The Tree 23
    Wind Songs 26
    Alone in the Wild 29
    Grabbin' Suckers 33
SUMMER 38
    I Wonder if They Will Remember 39
    Attack of the Midnight Bass 43
    I Want to be a Kid Again 46
    Night Lights 50
    Frog Hunters 53
    The Cabin 56
    Love You to God and Back 60
    Do Cows Chew Gum? 64
    A Never Ending Story 66
    The Last Cast 71
FALL 76
    The Old Barn 77
    Boone & Bo The Autumn of Their Years 81
    Cheap Meat 84
    Stump Sitting 88
    Everyone Needs a Special Place 90
    Deer Camp Characters 94
    A Family Tradition 99

Time Around a Campfire is Something Special 102
The Old Man in the Mirror 105
After the Fall – Saying Goodbye 109
WINTER 114
Deer Memories 115
Serendipity 119
Moon Walking 123
Sitting on a Big Flat Rock 126
What Not to do on a Cold Winter Day 129
My How Things Have Changed 133
Tracks in the Snow 137
Changing Lives 141
A Christmas Letter from Papaw 144
An Outdoors Legacy 147
About the Author 152

DEDICATION

For My God
My Wife
My Kids
My Grandkids
My Friends

ACKNOWLEDGMENTS

Without the support and encouragement of many people this book would have never been a reality. It is as much theirs as it is mine.

First of all, I want to thank God for his grace and for blessing me with the gift of putting my thoughts into words and words into stories. I especially want to thank Him for blessing me with a wonderful family and great friends.

My wife Maryann never gave up on me even though I gave her every reason to. She listened to my thoughts, read stories, offered advice and gave me encouragement. I love you more than the sky is big.

To our sons Daron, Kelly and Jeff, I'm so proud of you guys. I miss the times when you were little and wish I could go back and do some things different but I can't. Remember the good times and always know I love you.

To my daughter-in-law's LaVay and Lexi. Thank you for loving our sons and giving us the best grandkids in the world. I love you both.

To my grandkids Brittney, Hunter, Anna, Ty, and Sam you are all so special to me and grandma. You are a big part of this book just as you are a big part of our life. My granddaughter Anna did the art work you see in this book. Even though you have all grown up, we still need our hugs and to hear the words "I love you".

So many people have been helpful and supportive. My sister Peggie, Katie Mitchell, Bob and Barb Kipfer, Bob and Jean Rosenow, Lori Copeland, Sharon Kizziah-Holmes. I thank you all.

Thank you Bobby Whitehead, Lee Radcliff, Rance Burger, Tyler Schwartze, Tim Huffman, John Radzwilla and others for thinking my writing was good enough for your magazines and newspapers. Thanks David Gray for letting me write for your blog. Dave Barus you made my articles better than they were.

This book would have never happened if Johnny Morris had not come to me over 30 years ago and asked me to put together a radio show to be called Bass Pro Shops Outdoor World. Later came the same thing but in print. Thanks John. Love you buddy! I will always be grateful. You and Bass Pro Shops will always be a big part of who I am.

Finally, my style of writing is influenced greatly by two people whose books I have read many times. I still hear the voice of Charles Kuralt when I read the descriptive words he wrote. Many years ago I picked up a book called *Friendship Fires* by Sam Cook of Duluth, Minnesota. Sam doesn't know it but he was a great influence on my style of "writing from the heart" as well as what I feel in my soul when I am out in God's great outdoors.

Larry Whiteley

Springfield, MO

INTRODUCTION

I'm lucky to live in an area of the country where we experience the different seasons and I get to do that with family, friends and those special times alone in God's great outdoors.

Something would be missing in my life without the sounds of the spring peeper frogs, lightning bugs on a summer night, the autumn tapestry, animal tracks in a winter snow, an eagle flying in a bright blue sky or the beauty of a rainbow.

Changes in the seasons and all the special things that happen during those times, make up our lives. They add to it just as special times and special places, like seasonings added to a recipe, make life so good.

"Winter is an etching, spring a watercolor, summer an oil painting and autumn a mosaic of them all."

– Stanley Horowitz

SPRING

"Science has never drummed up quite as effective a tranquilizing agent as a sunny spring day."
— *W. Earl Hall*

THE EAGLE SEES THE ROUND RAINBOW

The bald eagle's role as our nations symbol goes back to 1782 when it was added to the Great Seal of the United States. The eagle was selected because of its great strength, stately looks, long life, and because it is native to North America. The design went on to appear on official documents, currency, flags, public buildings and other government-related items. The bald eagle became an American icon. To us as Americans, along with our flag, the bald eagle represents freedom and all that freedom stands for and is worth fighting for.

Since ancient times the bald eagle has been considered a sign of strength. Babylon, Egypt and the Roman legions all used the eagle as their standard, or symbol. Eagles figure prominently in the mythology of nearly every Native American tribe. In most Native cultures, eagles are considered medicine birds with impressive magical powers, and play a major role in their religious ceremonies.

In some of their legends, an eagle serves as a messenger between humans and the Creator. Eagle feathers were earned by Plains Indians as war honors and worn in their feathered head dresses. In some tribes today eagle feathers are still given to soldiers returning from war or people who have achieved a great accomplishment.

Eagles are also mentioned seventeen times in the Bible. My favorite is Isaiah 40:31, "but those who hope in the LORD will renew their strength. They will soar on wings like eagles; they will run and not grow weary, they will walk and not be faint."

In the wild, a Bald Eagle will live 30-35 years. A full-grown Bald Eagle has a wingspan up to 7-feet. They can fly up to 30 miles an hour and dive at 100 miles an hour. Eagles feed primarily on fish, supplemented by small mammals, waterfowl and carrion.

Bald Eagles mate for life, and an established pair will use the same nest for many years. Over time some nests become enormous and can reach a diameter of 9 feet and weigh as much as 2 tons. The female lays 2 or 3 eggs and both parents share incubation and guard them diligently against predators. While the chicks are small, the parents move about the nest with their talons balled up into fists to avoid harming them.

For such a powerful bird, the Bald Eagle emits surprisingly weak-sounding calls that are usually a series of high-pitched whistling or piping notes. The female may repeat a single, soft, high-pitched note that is said to be unlike any other calls in nature.

Fishermen who recognize the sound of an eagle usually stop fishing just to watch this majestic bird soaring in a bright blue sky. The bonus is when they dive from the sky to the water to do a little fishing themselves. Campers, hikers, canoers and kayakers are sometimes also treated to the sights and sounds of the eagle. It's a memory that stays with you forever.

Many years ago I was flying back home to Springfield, Missouri from Chicago in an old prop airplane. The plane flew very low all the way back. As I watched out the window I thought to myself, "this must be what an eagle sees as he fly's around". I pulled out a piece of paper and started writing the below poem and finished it before we landed.

The line about the round rainbow was added later and the title was changed after my wife and I were flying back from Florida. When we looked out the window of the plane we were amazed to see a round rainbow with the shadow of the airplane right in the middle of it.

Rainbows are created when the sun reflects off rain drops reflecting a multitude of colors. Most people don't realize that a rainbow gets its traditional semicircle shape from the horizon because we are only seeing half of it. When the same atmospheric conditions that create a rainbow are observed from an airplane or by an eagle, a rainbow is a full circle. A round rainbow is called a glory, which NASA defines as an optical phenomenon. To us, this glory was a sign that God was watching over us that day. He still is!

THE EAGLE SEES THE ROUND RAINBOW

*What must it be like to perch on a limb
in a tree on a mountain high?
Then look above and spread your wings
and fly into the sky.*

*The eagle sees the round rainbow
that has no beginning or end.
He sees the flatlands, hills and valleys
and places I've never been.*

*What must it be like to look below
at cloud shadows on the trees?
It must be wonderful
to be so wild and free.*

*The traffic on the roads must appear
like ants continually on the go.
Following straight and winding roads
to places only they know.*

*What must it be like to fly along the rivers
carving out the land?
Over ponds, lakes and oceans
all created by God's mighty hand.*

*The patchwork quilt of the fields below,
the prairies, the deserts, the plains.
How could you ever get tired of looking
when what you see is never the same?*

*What must it be like to fly over rows of houses,
giant factories, malls and other stuff?
For a majestic bird so used to nature's beauty
neon lights, billboards and concrete must be tough.*

*I wonder if tears come to an eagle's eyes
and they fall to the ground.
When he sees streams filled with trash instead of fish
and pollution all around.*

*What must it be like to fly above
when the seasons come and go?
To see the landscape turn from green to gold and red
to the white of a winter snow.*

*What must it be like to be an eagle
and soar way up high?
Oh the sights we would behold
if we could see through an eagle's eyes.*

SIGNS OF SPRING

Winter is losing its grip on the land and spring will soon be the victor. Signs of spring are beginning to appear everywhere. I'm ready for the change in seasons.

Last night as my wife and I were driving down the old road headed for another weekend at the cabin, we didn't talk much. Our minds and bodies were already gearing down and unwinding after another hectic, fast-paced week in the city.

Her thoughts were of curling up with a good book on the couch, visiting with neighbors, and going hiking. Mine were of making more hiking sticks, writing another chapter for this book, and wandering through the woods looking for turkey and deer signs.

Suddenly, the silence of our thoughts was broken as we drove by a roadside pond. Even with the windows rolled up we heard them. The chorus continued as we passed other ponds. We looked at each other and smiled. The peeper frogs are back.

To some, the first sign of spring is a robin in their yard, leaves starting to bud out, or flowers beginning to bloom. To cabin folks like us, it's the mating call of the peeper frog.

A single peeper frog is no bigger than your fingernail. One peeper by itself probably couldn't be heard even if you were standing right next to it. However, when hundreds of them blend their clear, birdlike "peeps" into a night-time chorus trying to woo a suitable mate, it's music to my ears.

We got to the cabin and hurriedly unloaded our things for the weekend. I fixed myself a drink and stepped out on the front porch. The peepers were still singing on the creek below and I could hear the musical trill of a toad joining in. The sounds of water cascading down our waterfall and water flowing across the riffles in the creek blended with frogs and toads and owls to form a symphony that was intoxicating.

The air was cool so I started to go back in when something caught my eye. The bats were back too! They beat their wings continually and with erratic, dodging, and short, soaring sweeps dive after flying insects that have also returned to the early night sky.

As I lay in bed that night, I could still hear the peepers as they continued to sing on the creek below the cabin. I drifted off to sleep listening to frog songs.

I awoke still thinking of peeper frogs and wondering what other signs of spring I might discover this day. Looking out the window, as night turned to morning, I sipped my first cup of coffee and watched the birds start coming into our feeders. Migratory birds were now joining our year-round residents. Even though I was inside, I could still hear them fill the air with their sounds of courtship. The drab goldfinches of winter are magically changing into the bright yellow of spring. Underneath the feeders, a chipmunk is busy stuffing his cheek pouches with seeds the birds dropped and then scurrying away to store them.

I know, too, that somewhere between here and South America, the tiny hummingbirds are on their way back to add to the mixture of bird life at the cabin.

SEASONS

The screen door creaks as I head outside with a cup of coffee in my hand and jacket draped across my arm. Birds scatter as a human enters their world. I set my cup of coffee down to put on my jacket when the rustling of dried leaves causes my heart to skip a beat and my eyes to look for a snake who has left his winter den early. Instead, a fence lizard scurries to a nearby rock to warm his bones in the early spring sun.

The barren trees of winter are starting to show signs of spring. As I look out through the treetops, I can see the white blossoms of the sarvisberry. They are always the first to bloom down here in the valley. It won't be long now until the redbud trees will burst forth with tiny purplish tinged blooms and the beautiful, showy dogwood will follow adding color to blue skies.

As I head down to the creek, I make a mental note of the location of all the trees that are the first to leaf out. These are buckeye trees that I go back to every fall to claim their bounty of buckeyes. Every visitor to our cabin has to leave with a buckeye in their pocket. It's good luck, you know.

I look for the tiny lime-green leaves of the maples but they're not ready yet. How many of you, as a kid, would toss the winged seed of a maple into the air and watch it twirl to the ground like a miniature helicopter? I still look forward to it.

I head off through the woods and see tiny wildflowers-- Dutchman's Breeches, Bloodroot and Mayapples, all pushing their way through the decay of the forest floor. The red of the Butterfly Weed, yellow of the Black-eyed Susan and blue of the Chicory will soon add more color.

I sit down on an old log to pause and enjoy this moment in time. My eyes catch the graceful, rippling flight of a Mourning Cloak butterfly as it emerges from its winter home in a tree cavity. A bee buzzes a wildflower. A spider is spinning his web on a nearby bush. A cricket walks up my pant leg.

The world is coming to life again and I'm thankful I'm here to enjoy it. All the sights, sounds and activities that are beginning to happen remind me that we humans weren't the only ones waiting for signs of spring.

GRANDPA'S OLD SHOTGUN

I sat in my turkey blind holding my grandpa's over one-hundred-year-old shotgun imagining him holding it waiting for a turkey. It was a 16-gauge Iver Johnson single shot, break-open model made in the late 1800s. I had been gifted the gun several years ago, but it had a broken firing pin, so it just sat in my gun safe.

One day I was looking for something in the safe. I pulled the old gun out, looked it over and wondered how Grandpa got the dents in the wooden stock or scratch along the barrel. I also wondered what kind of game he had shot with it to help feed his family.

I thought to myself how I would sure like to take a turkey with Grandpa's old gun. A local gunsmith fixed the firing pin for me and I took it out and sighted it in. I couldn't help but think of Grandpa as my hands wrapped around the old wooden stock.

On the other end of the property my grandson, Hunter, was in his blind also waiting and with one of my old shotguns. He was probably thinking about heading off to his first year of college at Kansas State University. It was his first time to hunt turkeys without his dad or papaw in the blind with him. That was a little hard for me, but I knew we had to let him grow up.

As I sat there in the darkness, the early light filtered through the trees and birds started singing their morning songs. I looked up at the fading stars and said a silent prayer thanking God for this special time in the outdoors that He created. I thanked Him for all the blessings He has given me and especially for a grandson that has grown up loving the outdoors as much as I do.

Matthew 21:22 says "And whatever you ask in prayer, you will receive, if you have faith." I asked God that if it be His will He might grant me the opportunity to take a turkey with the same gun my grandpa had also used to take turkeys so many years ago. I also asked Him to watch over my grandson, keep him safe and, if it wasn't asking too much, please put a turkey in front of him too.

The turkeys gobbled from everywhere all morning but wouldn't come into my calls. As great as I was sounding, if I was a gobbler I would sure come running to find the ladies.

I was about to give up when I saw a gobbler's head bobbing along a hundred yards away across the field from me. He wouldn't respond to my masterful calls or look at my amazing setup of decoys that should entice any old gobbler to come investigate. He just kept walking to the other end of the field over 300 yards away. I watched him through my binoculars as he crossed the road and went up into the woods.

I put down my calls, took off my face mask and gloves, poured a cup of coffee from my thermos, then set back and just shook my head. As I sat there, I heard a shot from across the property and hoped my grandson got his bird on his very first time turkey hunting by himself and doing all the calling. He must have sounded sexier than me.

As I prepared to pack up and leave to maybe try another area of the property, I looked up and saw that same turkey coming back out of the woods and angling across the field right toward me. I put my coffee down, hurriedly slipped my gloves and mask back on and got my grandpa's shotgun ready. To my amazement he just kept coming without stopping to gobble, looking at my decoys or me without having to make any kind of calls to entice him.

The gobbler never stopped until he got between my decoys and the blind, and then he turned his head and was looking right at me when Grandpa's shotgun ended the hunt.

I like to think Grandpa was with me at that moment and he was smiling down on his seventy-year-old grandson. I hoped my grandson might take a turkey someday with his great-grandpa's shotgun.

Oh, by the way, when I walked over to the turkey lying there in the wet grass, I broke open the shotgun to remove the shell, and I stared in amazement as I removed it from the gun. It was purple, the school colors of Kansas State University.

I have hunted turkeys many times in my life and I have never seen a gobbler do what that one did that day. Was it all just a coincidence? I don't think so!

God answers prayers if we just have the faith to ask. He answers small prayers and He answers big prayers. It may not be the answer we wanted or the answer we were looking for, but He does answer them. This day it was the answer I was hoping for, and I looked up and thanked Him then went to find my grandson.

The Sounds of Nature

There are some sounds in the great outdoors that you hear and they touch your soul. You don't have to see what made the sound because when you hear it you instantly see it in your mind. You may even hear them and see them as you read these words.

To some, the bugle of an elk is like that and so is the haunting howl of the wolf or the call of a loon. It might even be a cougar's throaty growl, or the gruff huff of a grizzly or black bear. Those of us who don't live where these animals live rarely, if ever, get to hear these sounds in the wild unless we travel to where they are, but if we do, they linger in our memories. Can you hear them?

Most of us have sounds in nature that stir us. A buck grunt in a November woods, the sound of a majestic eagle flying over a quiet lake or a turkey gobble on a spring morning. It could be the kingfisher's rattling call, as he flies up and down the creek, or a coyote yelp.

Maybe it's the quacking of ducks or honking of geese as they settle onto the water. The drumming sound of a woodpecker trying to attract a mate, the booming sounds of prairie chickens during their mating ritual and maybe the strange sound of a woodcock doing his sky dance trying to impress the ladies, too. Some of us hope that one day we will once again hear the sound of the bobwhite quail. Can you hear them?

Songbirds also add to nature's chorus. Chickadee's sing *"chick-a-dee-dee-dee"*, cardinal's join them with their *"purdy-purdy-purdy"* and the robin's whistling *"cheerup-cheery-cheerio-cheerup"* are joined by the tweets and whistles of all their friends. The squeal of a hawk can silence the bird music and get the squirrels barking an alarm to their buddies.

Owls ask us "who, who, who cooks for you?". Crows "caw, caw, caw" and then caw some more. The sound of peeper frogs or a whippoorwill means spring is finally here. The flapping sound of hummingbird wings and their distinctive chirp will soon follow. The rhythmic choruses of katydids can be so loud they drown out nearly all other sounds. Tree Crickets are known as the thermometer cricket because you can count the number of its high-pitched musical chirps in fifteen seconds and add forty to calculate the outdoor temperature in Fahrenheit. Believe me, it works!

The sounds of nature are everywhere in the wild if we just take the time to listen, and it's not just from the animals and birds. A rush of wind through the treetops, the rattling of dried fall leaves in a breeze and the sound of crunching leaves as something nears your secret hiding place. Booming thunder, the crack of lightning and rain dripping on a tent or the popping and crackling of a campfire. A stream tumbling over rocks and the soothing sounds of a waterfall small or big are music to the ears. To some it is the ocean waves crashing onto a sandy beach. To others it may be the "plip, plop, plip, plop" sound of a jitterbug gurgling across the water followed by the loud splash of a big bass rising out of the water to engulf it.

Nature sounds not only soothe our souls, but they are also good for our mind and body. Researchers say there is a scientific explanation for why sounds from nature have such a restorative effect on us. According to a study, they physically alter the connections in our brains to keep other thoughts out and the sounds even lower our heart rate. The exercise we get going to and from our listening places is an added benefit.

You're not likely to hear or, for that matter, see wildlife unless you force yourself to *just sit still*. We hike, we hunt, we fish, we camp, we canoe, we are continually on the *move when in* the great outdoors and not very quietly. We also carry with us the baggage of everyday worries, what's on the news, bills to be paid and work to be done.

You have to block all that out. Remaining still and quiet and actually paying attention to the sounds of nature is what is important. But that doesn't come easy. You can't just stop, listen for a few minutes and then move on. You have to settle down and tune into the sounds around you.

Those of us who sit in a treestand and a turkey or duck blind usually have no problem doing that because we have to if we want to be successful. If you just want to go out and listen to nature sounds though, I suggest you find a fallen tree, a stump or a big rock. Make a comfortable cushion of leaves, pine needles or take along some kind of pad and sit down. Now, don't do anything but relax. Don't let restlessness or thoughts of other matters creep back into your mind. Stay relaxed and breathe slow and easy. If you remain still the wildlife around you will forget you are even there. Soon enough the sounds of the wild will return.

The real art in listening to nature is not so much hearing the sounds of nature as it is identifying them. Listening carefully to nature sounds, and learning what makes that sound can help you begin to distinguish one sound from another and that gives you a greater appreciation for what you're hearing. The digital age has made it easier than ever to school yourself in nature sounds. Although this and other aids may be able to help, there's no substitute for firsthand experience. It's not just an ability to identify sounds, but also an understanding of their meanings that will come to you when you spend time listening carefully.

Yes, you can download and listen to nature sounds on your computer, tablet or smart phone. I listen to nature sounds accompanied by the melodic sounds of the native American flute as I drive down the road in my truck. If it is a cold, nasty, day not fit for man nor beast, I will put my headphones on and drift off to sleep listening to the sounds of nature. That is all good too, but it does not replace actually being out there in the great outdoors and being stirred by the sounds of nature that call us back to the wild. Can you hear them?

DAYDREAMS AND NIGHT DREAMS

The old gobbler is searching for me. I've done a good job making him think I'm a hen. He's literally tripping over his beard as he comes in looking for love. His bronze feathers shine in the light of the early morning sun, and the red, white and blue of his head stand out against the emerging spring greenery.

I give a soft putt with my mouth call and he comes in a little closer. To show me how handsome he is, he puffs out his body and spreads his tail feathers. My heart is pounding out of my chest as I stare down the barrel of my 12 gauge and slowly move to click off the safety. Suddenly there is a buzzing in my ear. What is that? The biggest gobbler I have ever seen disappears as I reach across my body to shut off the alarm clock.

My wife sleeps peacefully as I lay there for a moment trying to get the cobwebs and thoughts of turkeys out of my head. My feet hit the floor and my morning work-week ritual begins.

It's off to the kitchen to put on the coffee, a quick visit to the bathroom and then turn on the TV to catch the weather forecast. When I drink too many liquids before going to bed, the order of events sometimes changes. It can't be because I'm getting older.

It's back to the kitchen to pour my first cup of coffee, check the thermometer in the kitchen window and back to the TV just in time to see the local weather girl.

After thirty minutes of exercising, it's on to the bathroom again to shave, shower, brush my teeth, and get rid of the first cup of coffee. Back to the kitchen I go again for my second cup of coffee.

Now it's shirt, pants, socks, shoes and I'm dressed for work. I grab my briefcase and head for the kitchen again to fix my lunch. Before I head out the door to my truck, it's usually one more visit to the bathroom to get rid of the second cup of coffee. As I wash my hands, I look in the mirror and wonder if it really is because I'm getting older.

I stop by the bedroom to tell my wife I love her and get my morning kiss and then it's out the door and another morning routine has ended. As I start my truck, back out of the garage and head down the driveway, I wonder if I am the only one who does things in the same way, at the same time every morning. I think not.

As I drive down the road my eyes are always on the watch for deer at the forest edge. Maybe I'll see that fox again pouncing on a mouse in the field. That is if the red-tailed hawk doesn't beat him to it. Man, six road-kill skunks at the side of the road within two miles. That has to be a record! Around this curve is where I nearly always see turkeys. There they are! Six hens, a gobbler, and two Jake's. I wonder if that's the gobbler in my dream.

I'm sitting at a stoplight waiting for the green arrow and I see them. Geese flying in a V formation heading north. I wonder why we seem to notice them more when they're heading south for the winter rather than north for the summer?

I want to roll down my window and yell at the lady next to me, "Hey, look at the geese flying north! Do you know why they fly in a V formation?" She would think I was a crazy man so I think I'm better off keeping them to myself. Did I hear a goose honk? No, that's the guy behind me telling me the light's not going to get any greener. I make my turn and he passes me. Is he pointing at the geese in the sky too? If he is it's the wrong finger.

I exit on to the interstate highway filled with cars and trucks driven by people who have just finished their daily morning routine and are now off to work like I am. It's only a few miles before I will exit again, but this is a special time to me. Unlike those around me, I don't have the radio on listening to loud music or talk shows. This is my time for daydreaming.

My daydreaming each morning takes me to many places far from the busy highway. Sometimes I'm on my way to our cabin. I'm watching all the hummingbirds swarming like bees around the feeder or I'm down at the creek and I'm fighting that big, elusive smallmouth.

Other times, I'm heading north to the hunting cabin. You can't believe all the morel mushrooms and deer sheds I've found in my daydreams. I've also drawn my bow back on the biggest buck I've ever seen.

Daydreams have also taken me back to the mountains of Colorado, Montana, and Idaho. I'm there watching mountain goats working the salt licks, moose in the beaver ponds, elk in the meadows, and waterfalls caused by the melting snow.

I've also been to the mighty oceans and walked the sandy beaches with my wife looking for seashells, made sandcastles and snorkeled among the beautiful coral and colorful fish.

My daydreaming this day was of a special grandson and playing in the waterfall at the cabin, using toy road graders to make roads in the gravel bar, and fishing with him in the creek. Some daydreams come from your imagination, others from fond memories.

The clicking sound of my turn signal interrupts my

daydream and brings me back to reality. One more stoplight and I'll be at work. I pull into the parking lot, shut off the engine and take a deep breath. No time for daydreams here.

At the end of the workday, I will get back in my truck and head back down the same roads and I will daydream again on my way home. Daydreaming is an escape from worrying about the price of gas, work that needs to be done, or our inept politicians in Washington. It's too bad more people don't take time to daydream.

Yes, I'm a dreamer. Always have been, always will be. I enjoy my daydreaming, and tonight, I look forward to where my night dreams will take me. Maybe I'll get that old gobbler this time.

A Text from a Grandson

I was working on mounting a turkey fan in my workshop after a successful hunt this past spring when I heard the sound of a turkey hen. My smart phone makes that sound when I receive a text, and the sound of a turkey gobbling when someone is calling me. The text was from our youngest grandson, Sam, in Wisconsin that read, "Do you have a scope for the pellet gun? Oh yeah hi!"

Since Grandma and I don't get to see them as often as we would like, we tend to forget that he and his brother, Ty, are no longer the little boys that couldn't wait until we came to visit and were excited to see us when we did. They have grown up and are now teenagers who, if we are lucky, might respond back to our text or talk to us when we call. If we are really lucky, we might get a hug or at least a hand squeezed three times which means "I Love You" when we are there to visit.

They had outgrown the small air rifles we got them several years ago to shoot at targets on the farm. Our teenage grandsons still enjoy shooting at targets and had also got into trying to shoot pests like pine squirrels and chipmunks with Dad. The old air rifles didn't have a scope or the power to do what they wanted.

I texted back to Sam that I would do better than that. He and Ty would both be getting new air rifles that fit them better and would have a quality scope already on them. A text came back, "That would be nice. Thank you."

I had purchased a really good air rifle with scope for myself several years ago. This particular air rifle is well known for its quality, accuracy and power. I wanted one to help control the squirrels and black birds getting in our backyard bird feeders as well as rabbits in my garden.

I still use it and I still love it. It has a quality scope and a break over barrel that shoots .177 caliber pellets powerful enough to take down big groundhogs causing problems too. That's what I wanted to get for them.

For Sam I decided on what they call an "underlever" and offers some advantages over break-barrels. The design really helps accuracy. Also, the extra weight of the cocking lever is placed under the barrel and I felt that would make the rifle steadier in his hands when he shoots.

For Ty I went with what they call the next generation of air rifle technology. It is fast shooting, hard-hitting, quiet and comes with multiple features so he could customize the fit and feel. It does have a slight kick to it, but I knew he was big enough now he wouldn't notice it that much. It also had fiber-optic bars for the front sight that I knew he would like. I figured the rifled barrel teams along with the quality scope would deliver great accuracy for him. It also had a break-barrel action that would be easy for him to use.

Because they use air compression rather than the explosion of a powder cartridge means they will not only be saving money not having to buy ammunition, they would also not be breaking the law or disturbing neighbors.

I knew both rifles would last them a lifetime and someday they will be able to pass them on to their kids or grandkids long after I am gone.

Another great thing was that when they are ready to move up to hunting bigger game, they are already used to basic gun handling safety. I will be looking forward to getting another text someday that says, "Papaw, we want to start deer hunting so can we get a deer rifle now?" That will be a great opportunity for Grandma and me to make another investment in the lives of our grandkids.

A few weeks after getting their new air rifles, I heard that turkey hen talking again as another text from Sam popped up on my phone that said, "Thanks for the pellet guns". That was followed by one all grandparents want to hear "Love you!" A few days later I got another text that said, "Last night we got a squirrel." I smiled!

THE TREE

I was on my way to our cabin when I saw it. I am sure I've seen it lots of other times. It was just a glimpse as I drove by. Why did it bother me so much then? It was just a bulldozer knocking over a tree. That happens all the time in today's world. We have to have more convenience stores, banks, and shopping centers, don't we?

People have always cleared fields of trees. They did it to grow crops or raise cattle so they could feed their families. The trees were used for firewood to keep them warm. Now, they push over trees and just burn them to get rid of them. When the shopping centers are done, people take their families there to feed them or shop.

What really amazes me is to see developers clear the land of trees for a new housing complex and then name the streets after them. Then, people that buy the houses go to the local home improvement store or nursery to spend hundreds of dollars on small trees to plant in their yards that will take years to grow as big as those that were once there.

As I kept driving, I tried to think about what I needed to get done when I got to my cabin surrounded by the woods of the Mark Twain National Forest. I tried to listen to what the guy on the radio was saying. It didn't do any good. I kept seeing the bulldozer pushing over the tree. Why couldn't I get it out of my mind? It was just a tree.

Maybe it bothered me so much this time because I'm getting older and wiser. Well, older anyway. My mind took me back to when I was a kid growing up on the farm. I would spend all day wandering around in the woods. The trees hid me from all the Indians that were after me. I dodged their arrows as I ran from tree to tree. My imagination entertained me back then. I didn't need TV, video games or a smart phone. Thank God my kids grew up enjoying the woods. Now my grandkids are discovering the wonder of the woods, climbing trees, and carving initials.

Other days, I would climb up into a tree's comforting arms and soak in the wonder of the woods or just daydream. I can still remember the odd shape, a weird knot, and the feel of the bark on certain trees. I wonder if some of those trees are still there. I wonder if my initials are still carved in them.

As I got older, I would head to the woods with my dog, Bo, and my little single shot .22 bought with money I had earned. I still have that gun and the memories of knowing I only had one shot so I couldn't miss when that squirrel ran out on a limb. We didn't have a lot of extra money to be buying more .22 shells and that squirrel was supper.

I still enjoy hiking in the woods. I love the kaleidoscope of fall colors. I still climb trees but now it's to sit in a treestand waiting for a deer to walk by. My granddaughter poses for pictures on a grapevine swing. My grandson loves to hunt squirrels and deer now, too. I smile as I watch them, and I remember.

What was that the guy on the radio just said? "And he created the heavens and the earth." He created all the trees too didn't he?

It shouldn't be bothering me about seeing that tree pushed down. After all, I cut down trees, too, don't I? Their wood keeps our cabin warm during the cold of winter. They are also magically transformed into hiking sticks, candle holders, lamps, coat racks and lots of other things in my workshop.

I am wise enough to know that if your home is shaded by trees, your air conditioner won't run as much and you'll save money on electricity. You might even be able to open your windows and enjoy a fresh breeze. Cleaning the gutter, picking up limbs and raking leaves is a small price to pay.

Even my ten-year old grandson can tell you that the more trees you cut down, the less oxygen you have. Oxygen – you know the stuff that helps you breathe. I read somewhere that a single tree is valued at over $13,000 during its lifetime for the oxygen it provides. Multiply that times the number of trees in your yard, if you have trees in your yard.

Trees also are important to the wildlife that use them. Birds and squirrels build nests, turkeys roost, deer rub, woodpeckers peck. They feed on the nuts, berries, and insects they provide.

Fish and other aquatic species also rely on trees for shade along their watery homes. When they die and fall into the water, they provide fish habitat and safety from predators.

I pull into my cabin and a song is playing on the radio. As I listen, I'm not upset anymore. The words roll over in my mind as I look around at all the trees. "He grew the tree that he knew would be used to make the old, rugged cross." You see, that was the most important tree of all.

WIND SONGS

I don't remember the first time I heard the melodic sounds of a Native American flute, but the music still lingers in my soul. It is to most who hear it, an almost spiritual experience.

Legend has it that a woodpecker pecked holes in a cedar limb and gifted a young brave the first flute, but it wouldn't play. He had to first humble himself before it would sing. Since the heart of the cedar had been removed from the flute, it was his duty as a flute player to replace it with his own heart when he played.

I love to read about the time when America's Mountain Men traveled through the mountains and valleys of the west hunting and trapping animals for their fur. It was a tough life but I sometimes wish I had lived back then. They often heard the flute's haunting sounds and called the mystical music they made - "wind songs".

The Native American flute is the only melodic wind instrument belonging to the people of this continent and the only instrument indigenous exclusively to America. The oldest Native American flute is the Beltrami Native American flute. It was collected by the Italian explorer Giacomo Constantino Beltrami on a journey through present-day Minnesota in 1823.

Made mostly of cedar or river cane, they were used by many tribes for many different reasons. Some tribes used the flute for ceremonial purposes, in others young braves would use it to try and win the hand of their hopeful bride to be.

Mostly though, the flute was used to empty one's self of all the things which could not be expressed in words. North American flute music is a natural stress relief. In this crazy world we live in today, maybe we all need to learn to play a Native American flute or at least listen to their music to escape the craziness.

Since the first time I heard a Native American flute, something within me wanted to know more about it. How are they made? What gives them their beautiful sound? Can a musically challenged person, like myself, learn to play one?

I consider flutes as not just a musical instrument but also a work of art. Besides cedar and cane flutes they also make flutes of ash, maple, mahogany, blood wood, ebony, Alaskan yellow cedar and other woods from all over the world. Each has its own distinct sound and beauty when crafted by the hand of a master.

The flute is one of the easiest wind instruments to play. Minor tuning makes it easy because more notes go together than most contemporary instruments. A beginning flute player does not need to know conventional music when learning to play these instruments. It is a tool for self-expression. This simplicity allows non-trained individuals to be able to pick up the flute and make pleasing sounds within a matter of minutes. Master flute makers will tell you they have never sold a flute because a flute sells itself.

You don't have to play songs on a flute that everyone knows. Simply play what is in your heart. Look to a sunset or sunrise, the valleys and mountains, the streams and lakes, the wildlife and wild flowers. The world of nature contains countless songs. Look here for inspiration and play what you feel.

Native American flutes and lessons may be available in your area. You can also go online and order a flute, an instruction book, listen to flute music or order accessories. These special instruments, treated with care, will bring a lifetime of musical pleasure.

It is a beautiful early spring day. I sit on a special tree stump high on a hill overlooking a valley near the Mark Twain National Forest of southern Missouri. An eagle is flying in a bright blue sky.

I think of the Native Americans and how this was their land before the white man stole it from them. I think of how they took care of their land and tried to protect it from the white man's onslaught. I think of how they honored the game when they took its life to feed their family. I think of how they didn't waste any part of the animal and only took what they needed. They were the first conservationists. They were not savages – we were! They only fought to protect what was theirs.

My flute in hand, I play from my heart. It is an escape from this crazy world for just a little while, out in nature, away from it all. As I play, I also think of the Mountain Men listening to the haunting, mystical sounds of "wind songs" sweeping through the valley.

ALONE IN THE WILD

It's five a.m. on an April morning. I sit at my desk writing a blog article about going camping. My wife is still sleeping. The television is on so I can check the weather for the day. The weather forecast was a lot better than the news. It was about nothing but coronavirus. Sunny days, cool nights with a slight chance of rain. I turn the television off and go back to writing.

My days are spent following stay-at-home rules. There are always things to get done outside in the yard, garden or workshop. I had practiced social distancing and gone fishing a few times.

In a moment of absolute brilliance, I thought why not go do what I have been writing about. I rushed in and told my wife we should escape the pandemic for a couple of days and go camping.

She said she would rather stay around home but I should go enjoy myself. I stood there for a few seconds with thoughts rushing through my head of being alone for a few days in the outdoors. Alone in the wild.

I feigned disappointment and told her I would miss her. I packed all my clothes, camping gear and food in the truck. I also grabbed a couple of locater turkey calls.

As I drove down the driveway, I knew exactly where I was going. I would escape to a place that I was very familiar with. I had spent many years hunting deer and turkey there. I would go to an open area on top of a hill I had often thought would make a great place to camp. From there I could see for miles looking over forested hills and valleys but also with big open skies to enjoy. The creek in the valley below would be a bonus.

The stress and pressure from what was going on in the world with the coronavirus was gone as I drove up the hill. I pulled in by three trees that offered a great view. I just sat there for a moment. It was a totally different feeling than what I had been used to lately.

I pitched my tent and unloaded the truck. I got into my cooler for something to eat and drink then sat down in my camp chair to look around and take it all in. This is what I had come for.

The sun was warm. Sitting in the shade and with a little breeze, it was comfortable. I listened to bird songs. Crows were talking to each other. Buzzards circled in the bright blue sky. I looked up and said thank you to God for blessing me with this special moment in time. I also thanked Him for my family and not giving up on me.

My afternoon was spent fishing the creek in the valley. The water was cold as I waded and fished, but it felt good. I lost count of how many fish I caught. Nothing big but all fun. I tried skipping rocks and then just sat on the gravel bar looking for arrowheads and holey rocks. The sound of the flowing water was soothing. I took a nap.

When I woke up, the day was starting to fade so I drove back up the hill. The night skies were spectacular with thousands of twinkling stars. Coyotes howled and owls hooted. I did some hooting myself listening for turkey sounds from their roost. There were none. I stirred the campfire. The night cooled and my sleeping bag felt good.

I got up before the light came, stoked the fire and put on a pot of coffee. As the day started arriving, I was already out with my locater calls and binoculars scouting for turkeys. It wasn't long before I found where they were. I knew where I would be hunting when the season started. I went back to camp.

The smell of bacon sizzling in the skillet drifted through the morning air. A deer let me know they smelled it too. My second cup of coffee was as good as the first. Birds were singing again, and turkey gobbles echoed through the hills. Squirrels fussed at me because I was in their home.

The day found me secretly watching deer and turkey go about their day. I saw an eagle, a fox and a bobcat. Black bear roam these woods too. I didn't see one. I hiked around. I found wildflowers and morel mushrooms pushing their way through decaying leaves. I checked deer stands and pruned limbs and cleaned brush from around them. I even found a couple of shed antlers. I was enjoying my time alone in the wild.

Before I knew it, night was upon me again and the moon was big and bright. I sat around the campfire listening to night sounds and using my headlight to read "Friendship Fires" by Sam Cook. He doesn't know it, but his style of writing greatly influenced me. Friends Dave Barus, David Gray and Bobby Whitehead gave me the confidence I needed. They all shaped me into the writer I now am. I am using the gift God gave me.

My eyes are heavy from all my activities of the day, the dancing flames, a crackling fire and reading. I could hear thunder and see lightning in the distant hills. Tree frogs croaked and crickets chirped. Peaceful sleep came quickly.

Sometime during the night, I awoke to rain making music on my canvas tent. There is nothing more relaxing than that sound. I easily drifted back off to sleep.

When my eyes opened again, the sun was starting to shine through the trees. A light rain was still falling. When thunder rumbled, turkeys gobbled at the sound. I smiled. The sun glistened off the rain drops still clinging to the leaves and grass. I looked to the west and saw what I was looking for. A rainbow.

I sat there for a long time enjoying the beauty of the rainbow. Hundreds of purplish redbuds and white dogwood trees were all bloomed out painting the landscape. As much as I hated to leave, I missed my wife. It was time to go home to a different world. My time here will be re-lived in my daydreams and night dreams. It had been a wonderful escape from the pandemic. Alone in the wild.

GRABBIN' SUCKERS

Grabbin' suckers are nothing complicated, nothing new. Just ask the folks from Nixa, MO. It goes back to a time when families lived from the land. They raised pigs and fattened and butchered them. They milked a few cows by hand and drank the milk. They kept plenty of chickens for their eggs. When they wanted fried chicken for Sunday dinner, they would just grab one, cut off its head, pluck the feathers and fry it up in lard, made from the pig, on the old wood stove.

They always looked forward to April and May when sucker fish would school together in great numbers in the shallow shoals of local streams and rivers to spawn. Fish from the sucker family include yellow suckers, white suckers, blue suckers and redhorse. They were a special treat to the hard-working families, and they caught them any way they could.

In later years both farmers and city folk started using fishing rods with 20 to 30-pound test line, heavy sinkers and big treble hooks. A small white cloth was attached above the hooks, so they always knew where they were in the water. When they saw a sucker swim past the white marker, they would jerk hard and hope the hooks sunk into the fish.

Fishermen would stand on the gravel bars or elevate themselves on trees, rocks and even ladders to better see the fish in the water. Some even used stable flat-bottom boats. Polarized sunglasses became popular because they could better see the fish. There was no limit on the number of suckers you could keep back then.

Suckers are delicious but they are filled with tiny, thread like bones. To prepare them, they were scaled, then fileted leaving the skin attached. Cuts were then made through the filet about 1/8 inch apart to cut the tiny bones into small pieces. The filets are then covered in a flour and corn meal mixture making sure to get the mix down between the cuts and deep fried at 325 to 350 degrees for some of the best eating you will ever experience. Some locals canned or pickled sucker filet chunks to enjoy all year long.

Grabbin' suckers was so popular and such a longstanding local tradition it was suggested they should have a special weekend to celebrate this fish and the fishermen. The first Nixa Sucker Days was held in May 1957. Businesses closed and so did the school. Main street was lined with booths and games. Fishermen in their boats and floats of all kinds paraded down the street. There was musical entertainment, awards for the biggest sucker, a Sucker Day Queen was crowned, and of course, fried suckers were served along with all the fixins'. You could even have a bowl of 'sucker soup'.

I was an eleven-year-old Nixa boy at the time, and I loved it. I wanted to be a sucker grabber too someday. My uncle was Rex Harp who won many of the awards for biggest sucker fish. He was considered "King of the Sucker Grabbers" and always took his vacation when the suckers started their spawning runs.

When I was older, I worked to save money to buy everything I would need to be a sucker grabber. By then, I was married with kids and my weekends were spent grabbin' with friends. We enjoyed it because there was always plenty of action compared to regular fishing and having to wait and hope a fish took your bait.

When my sons got older, I started taking them. We have some great memories of sucker grabbin' together. By then, suckers were a twenty-fish limit per day instead of all you could catch. I fried a lot of suckers back then. The egg sac found in female suckers was a special treat when fried up just like I did the suckers.

For many years we went as a family to Nixa Sucker Days. It was a good time to see old friends and family, have fun, enjoy music and eat suckers. Sucker Days was always on the local news and was even featured one year on the national news.

As my sons and grandkids got older, we fished more for crappie, walleye and bass in the spring as well as went turkey hunting. The desire to go sucker grabbin' faded.

There doesn't seem to be as many folks sucker grabbin' anymore. Nixa Sucker Days has changed too. Most of the old-timers are gone. This year they will celebrate its 63rd year. It is now known as the *Nixa Sucker Days Music, Arts and Craft Festival.* Visitors can still get a chance to taste real fried suckers they say, along with other fried fish. There's still a parade and music, too, but it's mainly an arts and crafts festival now and not like the good ole' days.

I have fond memories of grabbin' suckers with friends and family. I remember great times spent at the old Sucker Days. My grabbin' rods are stored in the barn. Grabbin' suckers are on my bucket list. I keep telling myself I am going to go one more time. I am getting old. I need to do it while I still can.

A few years ago I was in Minnesota for an outdoor writer's conference. During an interview with the local Visitors Bureau, I asked what species of fish were in that area. They gave me a sheet showing and talking about all of them. They wanted to talk about the walleye, pike, crappie and yellow perch. I wanted to talk about the fish that was at the bottom of the list – suckers.

I asked them if people actually fished for them. They said, "No way! It's a trash fish. Nobody eats them. They sometimes catch them when fishing for other species and just throw them out for the eagles to eat or take them home and grind them up for fertilizer for their gardens."

I smiled and said, "Let me tell you a story about grabbin' suckers and a special day a town has every year in their honor." I even told them I would be willing to come back and teach them how to fish for them, show them how to cook them and pass out samples to the locals. I told them it could start a whole new fishing industry for them. They had no idea what they were missing. I'm still waiting for their call.

SUMMER

"A perfect summer day is when the sun is shining, the breeze is blowing, the birds are singing, and the lawn mower is broken."

— *James Dent*

I Wonder if They Will Remember

I'm sitting in a boat, watching the sun slowly rise and thinking about grandkids who once sat in boats and treestands with me and with Dad. They are teenagers and young adults now, and I wonder if they will remember all the memories.

When they were little they loved to come down to the cabin. Grandma and I have a lot of grandkid memories there at that place in the woods on the creek.

I wonder if they will remember campfires, poking around in it, watching the flames dance, and making s'mores. Will Ty and Sam remember their dad making them a wooden boat and sailing it down the creek or teaching them to skip rocks? Will Hunter remember using his toy truck and digger to make roads on the gravel bar? Sam loved to find beautiful rocks, holey rocks and beaver sticks too! Ty loved to shoot targets with a gun or bow. Anna loved riding ATV's and playing in the creek. They all loved to climb up on hay bales and grapevines.

Our grandkids were all oblivious to the complexities of life and excited about the little things like finding a turkey feather or a turtle shell, hunting squirrels, catching a fish or a crawdad, walking through a puddle, playing in the waterfall, snorkeling and Grandma hiding Easter eggs.

I wonder if they will remember, when it was time to go home, Grandma and Papaw would run out on the road to wave goodbye and blow them kisses. Will Ty, in his Daddy's arms, remember saying, "We're going home but we're coming back".

I hope they remember other memories besides at the cabin. Like helping Grandma bake cookies and Papaw make cinnamon rolls. Beating us adults in board games and video games. All the trips to Disney World and Florida beaches. I hope someday they will take their kids, too.

Will they remember feeling important when they were handed a trophy or ribbon from kung fu classes, bowling, basketball, baseball, gymnastics or showing cattle? Will they remember looking up in the stands and Mom and Dad and Grandparents were there for them?

Will they remember climbing up in Mom or Dad's or Grandma or Papaw's lap and having them read a book and then feeling proud when they could read to us? Will they remember Mom talking to them as she drove them to school or the tears in her eyes when they drove themselves, got on that big yellow school bus or went away to college?

Will they remember running and jumping in my arms, Dad patiently explaining how to do things, Grandma and Mom taking them shopping, having bad dreams or hearing a storm but knowing Mom and Dad were close by and would snuggle with them so they could go to sleep?

Hunter knew if Dad was too busy or too tired to play ball, he could call his papaw and I would be right over to play on that little indoor basketball goal. He was Michael Jordan and could really make those 6-pointers. Ty knew Dad or Papaw would play basketball with him. Sam knew Grandma would scratch his back and Papaw would squeeze his hand three times silently saying, "I love you". Anna knew Grandma would take her shopping and we would be there to watch her do her cheerleading and gymnastics. They all felt safe when Dad ran along beside them the first time they rode their bike without training wheels. Sammy wanted to know if his Grandma was proud of him when he did it.

I want them to experience things with their kids and grandkids like summer nights filled with fireworks or lightning bugs, catching a frog or turtle and giving them a name and tears run down their cheek when they escape, or they are told they have to let them go.

I want their kids to pick up a garden hose and squirt all the adults in sight or play on a slip 'n slide until their toes and fingers get all wrinkled. I want their kids to slide a snake down a sliding board and cut down the neighbor's tree with a hatchet.

I want their kids to dress up like Spider Man and make things out of Lego's. I want them to scratch their kids' back or lie down on the driveway again while they outline them in chalk just like their grandma did.

I want their kids to play snow football, build a snowman or a fort, and make a snow angel. I want their kids to do a cartwheel in the yard or on the beach. I want their kids to take cardboard and slide down the steps.

I want them to climb a tree and carve initials in it or sit up in it with their dad or papaw hunting for deer. I want them to go turkey hunting and walk down a lane holding hands with their dad.

I want them to be like Dad who lifted them up so they could put a ball in the basket or help them swing a big bat. I want them to jump on the trampoline or play in the pool with their kids and grandkids. I want them to toss around the football or play whiffle ball with their dad and papaw.

I want for them to always believe in the power of God, family, smiles, hugs, saying thank you, kind words, truth, justice, peace, dreams, and imagination. I want them to know that their kids will spell love "T.I.M.E". I want them to tell their kids, like I told them, that they can do anything they put their mind and heart to.

I want them to know that if they mess up in life that God is always there for them and so is Mom and Dad and Grandmas and Papaws even though we may already be in our heavenly home.

I want them to realize how much their lives would have been different if they did not have a mom and dad and papaws and grandmas that love them so much and were there for them. A lot of kids, like me and Grandma, didn't have that. I hope they say thank you, I love you and give us hugs as many times as they can before it's too late and they no longer can do that, and all that is left is the memories that I hope they will always remember.

ATTACK OF THE MIDNIGHT BASS

Pete Dye is 96-year's old. He still cooks for himself, does all his household chores and drives his car. No nursing home for Pete. In fact, somebody told me he still goes dancing. He has more energy than most men half his age.

The World War II veteran joined the Army in 1943 at the age of eighteen as an infantryman. He later trained as a paratrooper. Pete was in the Battle of the Bulge, the Battle of Bastogne and the Rhine River jump. He was with the 82nd Airborne Division during its occupation of Berlin and was prepared to go to Japan when the war ended in 1945. He was one of what is now called "The Greatest Generation".

Every Sunday morning you can find Pete sitting on an old wooden bench in the entry hall of the church we both attend. He has lots of stories to tell those who will take time to sit and listen. A lot of people do, including me.

His mind is still sharp, and he tells his stories in great detail. It is unimaginable to those of us who listen what these men went through. Many who died on the battlefields were in their twenties. Sometimes tears still come to his eyes when he thinks of buddies that didn't come home. Pete tells everyone God blessed him to live on, raise a family and have a good life. He and his wife, Wanda, were married for sixty-five years before she died in 2011. They had three boys and a girl. He has ten grandchildren and more than twenty great-grandchildren.

Each year we lose more of these men and women that sacrificed so much for us. We need to take the time to thank all our military who have served or are serving, but especially this generation before it's too late and they are all gone.

Pete knows I like to fish so he also enjoys telling me stories of fishing before hydroelectric dams created man-made lakes. A time before fiberglass bass boats that go faster than the truck that pulls them and cost more too. A time before mega-sized outdoor stores. A time when lures were mostly hand-made, rods were metal, line was braided cotton and reels were plain and simple.

Pete fished rivers, farm ponds and small natural lakes back then. If he caught something, he ate it, including bass. I like to listen to all of Pete's stories, but my favorite is a fish story that I have heard many times and I call it "The Attack of the Midnight Bass".

He always starts the story with "It's midnight on a hot summer day. The moon is full, bats are diving in the night sky, fog shrouds the lake. I am out fishing by myself in an old wooden boat. The night is filled with the sounds of crickets chirping, owls hooting and frogs croaking. I take my old bait caster and throw a topwater bait toward some bushes next to the bank. I let it settle then start reeling. It gurgles and wiggles back toward my old boat. Suddenly, a bass, a very big bass, attacks and thunders upward with the plug rattling in its jaw."

He always acts out the fighting of the bass until he got it in the boat. I love the expressions on his face while he's telling it. Sometimes Pete will add to the story with moans about backlashes of snarled braided line that required a lot of valuable time to untangle but he still got it in. Other times he might also include a story about a giant of a bass that shook loose from the hooks just as he was lifting him into the boat and it disappeared into the darkness of the water. Of course, the bass also gets bigger with each telling of the story. In fact, sometimes the bass was so big it was pulling the boat around. As far as I am concerned, he can tell it any way he wants to. I listen each time like I have never heard this story before.

When Pete goes home to heaven someday, I will miss him and hearing all his stories. There will be a lot of family and military buddies that will be glad to see him again when he gets there. I am sure they will all enjoy hearing Pete's stories as much as I do. Especially the one about "The Attack of the Midnight Bass".

I Want to be a Kid Again

The grandkids left a little while ago. They went back to the city with Mom and Dad. I'm sitting on the deck relaxing and missing them already. They love it down here at the cabin. There are a lot of grandkid memories here at this place in the woods, on the creek.

I wish I had their energy. In fact, I wish I had a lot of things kids have. I think I will officially give my resignation as an adult. I have decided I want to be a kid again.

I want to have a grandma and papaw that always greet me with a hug and kiss. Who let me do just about anything I want and will even do it with me. When it's time to go, I want to see them run out on the road to wave goodbye, blow me kisses, and tell me they love me. I want to say thank you and I love you to them again before it's too late and I can't.

I want to build a campfire, enjoy poking around in it, watching the flames dance, and make s'mores. I want Dad to make me a wooden boat and go sail it down the creek and use my truck and digger to make roads on the gravel bar. I want to find beautiful rocks and some with holes in them. Beaver sticks too!

I want to feel good because I can drive an ATV all by

myself but safe because Mom or Dad, or Grandma or Papaw is riding with me. I want to sing songs at the top of my voice as I ride and feel the comfort of big arms around me.

I want to be oblivious to the complexities of life and be overly excited about the little things again like finding a turkey feather or a turtle shell, hunting squirrels and picking apples.

I want to be excited when I catch a fish or a crawdad. I want to have fun walking through a puddle, learning to skip a rock or playing in the creek or waterfall. I want to go snorkeling. I want to laugh again as I push Papaw in the cold creek.

I want to look forward to helping Grandma bake cookies and pumpkin pies and Grandpa fry fish and make cinnamon rolls. I want to smile when I beat adults in board games and video games. I want to go to Disney World one more time.

I want to feel important when I'm handed a trophy or ribbon from karate classes, bowling, basketball, baseball, gymnastics or showing cows. I want to look up in the stands and Mom and Dad and my grandparents are there to see me. I want to know they are all there when I need them.

I want to climb up in Mom or Dad's or Grandma or Papaw's lap and have them read me a book and then feel proud when I can read it to them. Even though I didn't like it, I want Mom and Dad to help me with my homework again. I want Mom to talk to me again as she drives me to school and watch her cry when I drive myself or get on that big yellow school bus.

I want to know I can always run and jump in my papaw's arms and he'll catch me. I want Dad to patiently explain to me again how to do things. I want Grandma to take me shopping. I want to know that if I have bad dreams or hear a storm, Mom and Dad are close by and will snuggle with me so I can go to sleep.

I want to know if Dad's too busy or too tired to play ball

with me, I can call my papaw and he'll come right over. I want to ride my bike for the first time without training wheels and feel safe because Dad's running along right beside me and Grandma is proud of me.

I want to experience summer nights filled with shooting fireworks or catching lightning bugs and putting them in jars. I want to catch a frog or turtle, give them a name, and feel tears running down my cheek when they escape or Dad says I have to let them go.

I want to pick up a garden hose and squirt all the adults in sight or play on a slip 'n slide until my toes and fingers get all wrinkled. I want to slide a snake down my sliding board again and cut down the neighbor's tree with my dad's hatchet. I want to dress up like Spider Man again and make things out of Lego's. I want Grandma to scratch my back.

I want to play snow football, build a snowman or a fort, and make a snow angel. I want to play whiffle ball, do a cartwheel, and color in my coloring book. I want to take cardboard and slide down the steps. I want Grandma to lie down on the driveway again while I outline her in chalk.

I want to climb a tree, carve initials in it, shoot Indians from behind it and set in it with Dad or Papaw looking for deer. I want to go turkey hunting and walk down that lane holding hands with Dad again.

I want Dad to lift me up to put a ball in the basket and help me swing a big bat. I want to toss around the football with Dad or Papaw. I want to play on my little indoor basketball goal again. I really miss it!

I want to jump on the trampoline and my bed. I want to play sports for the fun of it. I want to believe, like Papaw told me, that I can do anything I put my mind to. Heck, I'd even look forward to going to school again.

I don't want to have to get up and go to work every day. I don't want to worry about budgets, deadlines, computer crashes, or go to meetings. I don't want to pay taxes or insurance anymore. I don't want to watch, read, or hear the news anymore either. It's all bad anyway.

I don't want to hear gossip. I don't want to worry about how much I have in my savings account. I don't want to worry if I did something or said something to offend anyone. I don't want to worry about my kids or grandkids, but I do want to be there for them and help them when I can.

I want to believe in the power of God, family, smiles, hugs, saying thank you, kind words, truth, justice, peace, dreams, and imagination. I want all parents and grandparents to know that kids spell love "T.I.M.E".

Okay, that's it! I'm through talking. You can have my checkbook, credit card, bank statement, house keys, car keys, or whatever else you want. I won't need them anymore. And, oh yeah, here's my resignation. I am no longer an adult.

So, tag you're it. I'm going outside. I saw a mud puddle I want to jump in.

NIGHT LIGHTS

The warm early summer day is ending. The bright orange sun slowly begins sinking to the earth. It's been a long, hectic day at work, and I step outside to begin winding down. I love watching sunsets and sunrises.

A lone whippoorwill calls from the nearby woods testing the silence and is answered by another down in the valley. Tall fluffy clouds gather on the horizon. The bottom layer lights up in varying shades of pink and orange like a painter mixing colors on his palette. Frogs begin their nighttime chorus and bats are diving for insects in the fading night sky.

As the darkness slowly settles, I see it. A tiny twinkling orb. First one and then another until suddenly the summer night is bombarded by a myriad of twinkling lights. I sit down on the front porch to watch the performance.

Gazing at the slowly pulsating lights, I travel back sixty years to Grandma and Grandpa's farm. As the adults sit around talking, we kids ran about capturing these jewel green sparks that pierced the dark and put them in Mason jars with holes punched in the lids. It was a magical time racing about filling your jar. Our eyes twinkled as much as the stars, and laughter pierced the silent night. I wonder how many other adults are outside like me right now and feel the stirring pleasures of childhood.

My mind also wanders to a special time one summer at our cabin. An approaching storm was playing music on our wind chimes awakening me from a deep sleep. The alarm clock by the bed told my sleepy head it was 2:30 a.m. as my feet hit the floor to go check out what was happening. I walked through the dark cabin and looked out the windows into the night. The blinking lights of fireflies were everywhere. This night though, they seemed much bigger than normal tiny fireflies. It was almost as if the window I was looking out was a big magnifying glass and I was seeing the insects much bigger than they really were.

I stood there in wide-eyed amazement as I watched them. They were high in the trees, they were down by the creek, they were up by the road, they were way down in the valley. How could I see them that far away? Maybe the sky was just darker than usual that night causing their lights to shine brighter. Maybe they were brighter because they were really trying hard to impress their lady friends. At the time I didn't really care what the answer was, I was just enjoying the show.

As the storm approached closer, lightning lit up the dark sky. It wasn't streaks of lightning though; it was more like bursts of light. It was like there were now gigantic lightning bugs joining in with the smaller ones to add to this special night.

I don't know how long I sat there watching but eventually the rains came, the lights went out, and I went back to bed. I lay there listening to the rain on the roof and grateful the storm had awakened me. I drifted off to sleep thinking of fireflies.

The neighbor's dog barks and my wandering mind takes me back to my front porch again. I'm thinking how I took a nail and punched holes in the lid and put them on jars for my kids. I hope they, too, have good memories of summer nights and twinkling lights. Grandkids are now learning to enjoy this age-old mysterious performance but instead of jars they use plastic firefly houses. Kids need fireflies more than they need television and computers, and so do adults.

As if saying goodnight, the tiny sparks blinked off one by one. I get up from the porch and head for the garage. I'm looking for a 65-year-old Mason jar with holes in the lid.

Frog Hunters

My dad and his friend, Gene, were returning from a very successful nighttime frog-gigging trip when Dad's truck's headlights went out. Looking all over the truck they couldn't find a fuse but lucky for them they found a .22 caliber bullet that fit perfectly into the fuse box and the headlights came on.

After traveling approximately twenty miles and just before crossing a bridge, the bullet apparently overheated. Unlucky for them it discharged and struck Dad in the right testicle. The truck swerved and hit a tree. Dad suffered minor cuts but did require surgery to repair the other wound. Gene suffered a broken rib and was treated and released.

Gene told everyone, "Thank God we weren't on that bridge when he shot his intimate parts off or we might have died." When Mom was notified of the wreck, instead of asking about poor Dad's wounds and where she could go see him, she asked, "Where's the frog legs, I need to get them in the refrigerator?"

Now, it's a good thing I was already born when this happened or I wouldn't have been here to try gigging frogs for myself. Back in my teenage years every boy learned how to gig frogs, cut their legs off and bring them home for their mom to fry up for the whole family. Of course, we were poor back then so Mom was good at frying up about anything I brought home, but she dearly loved frog legs.

It was a warm summer day and my friend, Frankie, and I decided it was time for us to be real men and go hunt frogs. We found an old burlap sack to put the frogs in so we wouldn't have to put wiggling, bloody frogs in our jean pockets.

Dad let us use his frog gigs since he had not gone frog hunting for years. I guess it was hard for him to gig frogs with one testicle. Mom was very excited at the possibilities of me and Frankie getting a bunch of big bullfrogs so she could fry up a bunch of frog legs again.

Dad also let us use his flashlight and gave us some tips and advice since we knew nothing about this great sport. One of the first things he said was to be alert for snakes. We were assured snakes found at a pond were harmless but a water moccasin snake, which is poisonous, could possibly be visiting the pond.

Frankie and I thought about that a lot on our way to the pond where we had gotten permission for us to hunt frogs. Some people may think there is not much excitement when hunting frogs. That's not true. Walking into a large spider web in the dark can cause anybody's heart to start beating really fast. When you realize you have walked into a large spider web, you then start wondering where the big spider is and screams escape from your mouth.

We knew there were cattle in a couple of the fields, but we didn't know which fields. Typically, cows are gentle animals. Cows are curious, and for their size it's amazing how quietly they can walk up to you in the dark. Your heart can start beating very fast when you turn and see glowing eyes about four or five feet above the ground staring at you. It gets more exciting when you realize the eyes belong to a 1,500 pound or larger cow that could quickly decide it didn't like you.

Thankfully there were no cows in the field on the way to the pond, but another cow danger in a dark field is a fresh cow pie. When you step in a fresh one, you can slip and fall. You do not smell very good when covered with cow pies.

Tripping over a stick, stepping into a hole or small ditch, slipping in mud, walking into a low hanging branch, etc. can get your heart beating fast too. This can be very disruptive to the frog hunt. Any of these exciting activities often cause a person to yell, curse or both. This type of sound is not normally heard around a pond. Anytime an unusual noise is heard by frogs, it will cause frogs to jump into the water.

Dad's flashlight was really good for spotting the light reflecting off the frog's eyes or any other animal's eyes the light found. A frog will freeze when light is shined in its eyes. This enables the hunter to get close to the frog before spearing the frog. It becomes tricky walking toward the frog and keeping the light shining in the frog's eyes. Moving the light beam off the frog's eyes can cause it to jump. Walking while holding a light beam on a frog's eyes is similar to walking while balancing a book on your head. It can be done but with a lot of practice.

We had numerous misses trying to gig a frog that night. We said words that if our mom's had heard them, they would have washed our mouths out with soap. We fell in the water and our arms and faces were scratched up by attacking limbs. Did I mention all the mosquito bites? We left the pond that night wiser in the knowledge of frog gigging but with no frogs. I dreaded facing Mom with no frogs. She loves her frog legs. Just ask Dad.

THE CABIN

When I was younger man I used to dream of having a cabin in the woods. Not a lake house sitting among rows of other houses overlooking a man-made body of water. Just a simple, secluded cabin nestled among the forest trees.

My grandma used to tell me if I dreamed long enough and worked hard enough my dreams would come true. Grandma was right! Thanks to a little luck and maybe some help from the man upstairs my wife and I found that cabin.

It sat upon a rock bluff overlooking a creek. Just like my dreams, it was nestled among cedars and hardwoods and a scattering of pines. The trees keep it hidden from view of the few cars that traveled the gravel road nearby. The trees offered shade and protection from the summer's sun and winter's cold winds.

The cabin was built in the 1950's over what looked like the foundation of another cabin. I always liked to think the old foundation was a trapper's cabin from the late 1800's. The cabin was small. It had a kitchen, bathroom, closet and a combined bedroom and living room along with a big deck. Water was pumped up from the creek.

Over the years it was remodeled inside and out with log and pine siding. We liked the original cabin, but it just wasn't big enough for family gatherings or hosting friends.

A well was drilled which gave us the luxury of taking baths in an old claw foot tub instead of using a garden hose or the cold creek. The deck was also expanded to extend over the bluff and around the side.

We built a small barn for storage and to serve as a workshop for making things. Later we built an even bigger barn for ATVs and a bigger storage and workshop area.

Inside the cabin was a little wood stove sitting in a corner. It warmed the cabin on cold winter days. On one wall antique snowshoes hung on both sides of a pair of moose antlers. A mounted deer head, turkey fans, a pheasant, ducks, trout, bass and a big musky covered other walls along with fox, beaver and raccoon pelts. Each had a special memory and a story. A 4-foot timber rattlesnake skin got a lot of attention from first time guests.

My old muzzleloader and powder horn hung above the front door and rusted traps sat around on shelves. On a wall near the bathroom door hung a wood carving of a mountain man. My wife bought it for me for my birthday because she knew I loved that period in history.

Deer antlers, turtle shells, feathers, buckeyes, rocks, bird nests and other nature things could be found everywhere you looked. Most had been picked up by grandkids while on cabin adventures. They were mixed in with antique duck decoys, jars and dishes that were my wife's special touch.

Most noticeable though were the pictures of our kids and grandkids hung with loving care on walls and sitting around everywhere; Pictures of them with turkey and deer and fish or just having a good time at the cabin.

Looking through the windows we would see birds of all kinds coming to our many feeders. April thru October was hummingbird time, and I don't mean just a few. Hundreds at a time. It was a sight that thrilled everyone that visited the cabin.

The front porch and big deck were a great place to watch

squirrels playing in the woods, deer walking through, butterflies landing on wildflowers, or to see bats diving for insects in a summer's night sky.

From there you could also hear a waterfall that cascaded down what we called Dogwood Mountain. You could listen to the sounds of the creek as it flowed across the riffles below. You could see the kingfisher swooping above the water and hear crows calling down in the valley.

The fire pit off the deck out back was where grandkids roasted marshmallows and shared time with Papaw. It was a place to watch the flames dance and flicker as the worries and stress of the work week melted away. Fish fries, cookouts and fellowship all took place around the fire pit.

Grandkids loved going fishing, hunting squirrels, swimming, snorkeling, catching crawdads, skipping rocks, playing in the gravel or waterfall, finding feathers, wading in the creek and riding ATVs at the cabin. Christmas at the cabin with family was a special time in a special place.

Good neighbors also helped make the cabin what it was. Bob and Barb, Wayne and Jane, Annie and Winnie, Judge John, Doug and Kim, Sheila, Deke and others. They loved the valley and nature as much as we did. With them we shared hiking trails, ATV rides, campfires and pieces of our lives.

Spring at the cabin featured redbuds, dogwoods and wildflowers. The sounds of spring were peeper frogs and whippoorwills. Summer was fishing, swimming, relaxing or playing in the creek. Fall brought a kaleidoscope of color, looking for buckeyes, hiking, deer hunting and cutting wood for the cold months ahead. Winter meant books by the fire, making new hiking trails, and hiking in the snow.

So many memories were made in the twenty years we owned it. We loved that cabin! The time came when kids and grandkids got older and didn't come as much. They were busy with other things. As my wife and I got older it was harder to do all the upkeep a cabin in the woods required of us. It was a hard decision, but we decided it was time to sell it if our kids didn't want it, and they didn't.

We decided to try to find a couple with young kids who would love the cabin as much as we did and had time to make memories like we had done. The man upstairs blessed us again with finding that special family.

The cabin will live on with them. My wife and I and our family will always have all the memories from this special place. That is something that never goes away.

You see, a cabin is more than just a cabin. It is a living structure with a soul of memories and dreams. It is a place to get away, to share with others and to share fragments of one's life with the nature that surrounds them at the cabin.

LOVE YOU TO GOD AND BACK

I am looking at a picture of a three-year-old little girl sitting on my lap at our cabin. As I look at other pictures of her my mind travels back to special times we have had together. I am proud of the young adult she is, but I miss that little girl too. Grandma and I have always told her "We love you to God and back".

I look at pictures of her playing in the swimming hole at the creek riding in a tire tube through the little rapids into the waiting arms of Dad or Papaw. I look at pictures of her snorkeling and touching the fish that Dad helped her catch.

There was the time she and I carved a Halloween pumpkin with my Dremel tool. We both had strings of pumpkin all over us from head to toe. We also carved deer tracks and other things into rocks and wood with that Dremel tool.

One of her favorite things to do was helping Papaw make his famous cinnamon roll along with her brother. When we were finished, they both would have flour, cinnamon and butter all over them. Grandma would help clean up the mess and we would all enjoy the delicious treat.

No matter where she went, she always brought her favorite "blankies". They were soft throws to wrap up in because they "smelled like home". I bet she still has some in the house she and her brother share at college. She loves her "Bubby" and they take care of each other.

We would sing "Zip-a-Dee-Doo-Dah" and other songs as we rode the ATV together down the cabin road. Her hair was always getting caught in her papaw's beard and she would smile. I love her smile!

One day she was bored so I pulled the ATV over by a field of wild thistle with butterflies all over them. I handed her my camera and told her to take pictures of them. She liked doing that and became so good at it she has won several awards for her photography work.

We saw wildlife, found deer antlers and lots of other neat stuff as we rode. Later she and Dad would ride the ATV on the cabin roads with a special dog named Memphis. He loved to ride the ATV too!

She loved it when her cousins, Ty and Sammy, would come to visit from Wisconsin. They climbed grapevines and hay bales, played in the creek and the waterfall, rode the ATV's, shot BB guns, cooked smore's over the fire and had lots of fun.

We would all go to visit them in Wisconsin too. Sometimes she would stay there with me and Grandma when Mom, Dad and brother would go back home. The first night she would always not feel very good. She would say she had a leg ache, but Papaw knew it was really a heart ache because she missed Mom and Dad. I would hold her close and tell her it would be all right.

As she got older she started going deer hunting with us. She and Dad spent a lot of great time hunting together while I hunted with her brother. Dad taught her well and has proudly watched her harvest several deer. Dad and I both agree that opening weekend of deer season isn't nearly as much fun anymore with both of them away at college and not there with us.

Her "Bubby" helped her get her first turkey and now

he's teaching her to hunt ducks. She still fishes too and will sometimes catch fish when the rest of us aren't. I think she just likes being on the water with her family.

Grandma and I have always said that Anna can do anything she sets her mind to. In high school she made the varsity cheerleading team as a freshman, made herself into an excellent student with good grades, and won awards for her art and photography. I have no doubt when she graduates college she will go on to make her dreams come true.

A lot of who she is as a person came from having a great mom and dad, grandma's and papaw's and a brother that were always there for her. She also knows she is a child of God and He is always there for her too.

At college she has her "baby", a rescue dog named Max. Her brother, his fiancé, Molly, and their dogs, Maverick and Willie, are there for her too. Mom and Dad visit often, and Grandma and I make the trip sometimes, too, when the grandkids need something.

Someday a young man will come along. I am betting that Dad will have a talk with him and her "Bubby" will be sure and show him pictures of her with deer, turkey and shooting pistols just to make sure he is the right one. If he is, they will start a family of their own and make memories of their own.

I hope she tells him and her kids stories of swimming holes and cinnamon rolls, ATV rides and hunting deer with Dad. I hope she tells them of trips with Mom to Florida beaches and trips to Wisconsin. I hope she tells them of special times at Silver Dollar City and Disney World. I hope she tells them of shopping with Grandma and tells them about her God who loves them too.

A tear runs down my cheek and falls onto one of the pictures. I quickly wipe it off and start putting them all away. Thanks "Sis" for all the memories.

I hope all of you who read this are making lots of

memories with your kids and grandkids. Remember, kids spell love T-I-M-E. Tell them you love them to God and back!

DO COWS CHEW GUM?

We were driving back home with two of our grandkids. Brittney and Hunter were in the back seat not saying much since they were a little tired from all the digging in the gravel and playing at the creek down at the cabin. Hunter had been chewing gum that had lost its flavor, so he asked Grandma to take it. She rolled down the window and threw it out. As we kept driving he was staring out the window and watching cows in the field as they ate grass. This then two-year old kept studying the cows and what they were doing until finally he asked, "Grandma, do cows chew gum?"

That's just one of many memories my wife and I have of grandkids. We cherish every one of them.

Sometimes the memories come from pictures. Brittney in her little strawberry dress or smiling holding the biggest fish she had ever caught. Hunter riding the ATV with Papaw or with his daddy walking down the lane to go turkey hunting. Anna swimming in the creek or fishing with her daddy.

Sometimes the memories come from watching videos or seeing pictures of when they were babies, Christmas around the tree, gymnastics, dance recitals, reading them books, soccer, baseball and basketball games.

Sometimes you wonder if one day they, too, will have

memories. Will Brandon remember fishing the creek with Hunter, snorkeling with Grandpa in the swimming hole and wishing he could stay there forever? Will Corey remember making pine cone feeders for the birds for Christmas and realizing this was a lot more fun than watching TV or playing video games? Will Brittney remember rubbing Grandpa's bald head and pinching his ear lobes or sitting on the roof of the house with Grandpa and telling him things she wouldn't or couldn't tell anybody else? Will Anna remember shopping trips and Grandma buying her things or teasing Papaw and not giving him kisses? Will Hunter remember Papaw playing basketball, football and baseball with him or spending time at the cabin with just him and Papaw and Grandma? I hope they will.

Sometimes just words bring back the memories of grandkids and good times. I'll say to my wife in a whispered voice, "You want to watch one of my movies?" and she will instantly be taken back in time to Hunter asking that question with wide eyes at three a.m. one morning at the cabin. One time I told Anna "I love you more than the sky is big" and she said "really" and put her head on my shoulder and squeezed my neck. Grandma and I now use that phrase to each other.

Our newest grandkids are Ty and Sam. Grandma and I look forward to making lots of special memories with them, too. Like the rest of our grandkids, they will also spend lots of fun times at the cabin. Ty memories have already started. Giving us kisses and hugs over the phone, reading books to him, and little hands under the bathroom door.

Sam memories are still to come.

My wife and I still drive by those fields heading home from the cabin, and sometimes I look over at her and ask "Grandma, do cows chew gum?" and she smiles.

A NEVER ENDING STORY

Bob and Barb Kipfer are two people with hearts as big as the outdoors they love. For those that know them, it is hard to think of one without thinking of the other. They are husband and wife, but they are more than that. They are friends, they are a team, they are partners in life. A life well-lived.

The first chapter in their book of life begins at Kansas University Medical Center. Bob was a medical student in his first year of patient care on the hospital wards. Barb had just arrived as a newly graduated nurse on her first job.

During his daily classwork around the hospital, Bob took special notice of Barb. One-day, Bob saw her going into a room where nurses went to dump bed pans. He followed her in, closed the door and asked her out on a date. He thought he might get dumped, too, but she said yes. They were married September 4th, 1965 and another chapter in their life had begun.

Two years later Bob received his draft notice and served with the infantry in Viet Nam as a battalion field surgeon which meant he traveled into battle with the troops and worked in field hospitals. Barb continued nursing back in Kansas and caring for their newborn son Mark while hoping Bob would make it back home to them. I am sure there were times when Bob wondered the same thing. Like most Viet Nam veterans, he doesn't talk much about that time in his life.

Needless to say, he did make it home to his family after his tour of duty ended. They settled down to a somewhat normal life during four years of his residency at the Mayo clinic. Their family also grew with the birth of daughter Amy. Life was busy, life was good.

In 1973 Bob and Barb and the kids moved to Springfield, MO to start a new chapter in their lives. Bob practiced Gastroenterology and Internal Medicine at a local hospital. Barb started teaching at a school of nursing. They bought a home and moved into an urban neighborhood where they still live today.

Their lives were busy, but they managed to find time to go fishing, canoeing, kayaking and sailing on weekends. They played tennis. They traveled. They made lots of friends at work, in their neighborhood and through social activities. One of those friends owned land with a cabin in the Ozarks hills of southern Missouri. Bob and Barb visited often, and they started looking for land of their own.

That search led them to land with a clear flowing creek running through a beautiful valley with forested hills and lots of wildlife. They fell in love with this special place and another chapter was to be written.

Bob and Barb continued to work at their medical jobs during the week and stayed at their home in town. Unless they were traveling to places all over the world, visiting their kids and grandkids in other states or going to social events, they were at their valley on weekends.

Ten years after buying the property Bob decided it was time for another chapter to be written. He was now working in medical administration in addition to his medical practice, but having more fun on their property, and retired. He gave up tennis for a chainsaw and a tractor down in the valley. Barb waited two more years before retiring just to make sure Bob was house broke.

Retirement started another chapter to their story. During their time they spent in the valley they started working with the Missouri Department of Conservation to clear trees to bring back glades that were once there. They also worked with the department to plant trees for bank stabilization to protect the stream and their land. They even planted over 2,000 tree seedlings themselves for the same purpose. This all sparked their interest in conservation and fed their desire to conserve and protect this special place.

Their transformation from medical professionals to dedicated conservationists and conservation educators is an amazing chapter in their book of life. It's about how their love for conservation grew and changed not just their lives, but changed and touched the lives of so many others. More than they will ever know.

They became involved with the Springfield Plateau of Missouri Master Naturalist. Bob writes an informative blog for the group. Barb represents them on the Grow Native Board and leads educational tours of their urban yard in Springfield where she has planted over 100 native species of plants. She even made a video tour of what she has done to be used for virtual education.

Barb spends a lot of her time in the valley trying to rid their land of any kind of invasive species of plants not native to the area. They have restored warm season native grass fields and work at endangered species protection. They collect native butterflies, raise moths, and volunteer at special events at the Butterfly House, as well as hosting mothing events at their property.

They implemented a forest stewardship plan for their property. It is now a certified Tree Farm. They were named State Tree Farmers of the Year in 2015 for all the work they have done with timber stand improvements and even hosted a Missouri Tree Farm Conference.

Their land in the valley has grown to 400 acres and

includes another cabin that came with their land additions. The valley and the cabin are used by college students for stream ecology studies. The Audubon Society has access for bird counts and education. They have hosted Missouri Department of Conservation tours, a black bear study, Boy Scout activities, wildlife studies of plant and animal species, wild mushrooms studies, as well as field trips for groups doing plant and wildlife identification. Their land is open to other conservation minded groups for retreats and field trips for ecology, woodland management and stream education.

They were named the 2017 Conservationist of the Year by the Conservation Federation of Missouri. I would bet if you asked them what they have enjoyed doing most of all the things they have done, it would be their work with the public schools WOLF program. They teach fifth graders in weekly classroom sessions as well as hosting the kids in their valley for educational classes several times a year. Bob and Barb have made a profound impact for conservation in the lives of all the kids they have taught. The kids love them and will never forget Bob and Barb. This world could use more people like the Kipfers. Their impact for conservation has been huge.

One of these days, I hope in the far distant future, Bob and Barb will no longer be able to manage their land. When that time comes, they have donated it to Missouri State University under a protected agreement to sustain the natural ecology of the valley and use it to educate students who will be our future conservationists and conservation educators.

When Bob and Barb are gone, their ashes will be added to the old cemetery in the valley they loved. Their passion for conservation will continue on through these students, the Wolf School kids and all the other lives that have been impacted by these two people. It will not be the final chapter of their book of life though. Their story will go on through all the lives they have touched. Those people will pass on their passion for conservation. The Bob and Barb story will continue. That makes this a never-ending story.

THE LAST CAST

He was alone on the lake. The sunrise was breathtaking. He had seen lots of sunrises but none this beautiful. His first cast landed near some bushes. He felt the thump and set the hook. The largemouth came out of the water trying to shake the bait. It fought hard but soon tired. He gently lifted it from the water, smiled and released it.

There would be many more fish that morning. One was the biggest smallmouth he had ever caught in all his years of fishing. The sunlight glistened off its bronze body. He managed to take a selfie of him and the fish. As he hit send on his smart phone, he smiled. A son texted back, "Nice one, Dad." Another son replied, "Good fish, old man!" A grandson asked, "What did you catch it on?" His wife texted, "Are you doing O.K. and how are you feeling?" He smiled and texted back each of them with only the words "I love you" and then went back to fishing.

It suddenly occurred to him that he had not heard or seen another boat all morning. Kind of felt like he was fishing on his own private lake. He heard crows, ducks and geese. He saw deer and turkey at the water's edge. Birds were flittering around everywhere and singing their songs. A hummingbird even came buzzing by thinking he was a big flower. He said to himself, "Is this what heaven will be like for a fisherman like me?" He smiled again.

The afternoon sun was high and hot. He motored into a shaded cove and shut off the engine. The slight breeze felt good there in the shade. He tied the boat to a tree, sat back and relaxed. Thoughts of the first fish he ever caught went through his mind. He saw the bobber, the worm, his cane pole. He felt the little perch squirming in his hand. The special feeling he had that day, alone on that creek, was unlike any other. He was hooked. It was the first of many fish he would catch in his lifetime.

As he stretched out in the boat, more memories flooded his mind.

He wished his dad would have taken him fishing, but he didn't. He thought of times he took his sons fishing. He recalled the look on their faces when they caught their first fish. He wished he hadn't been so busy trying to make a living and would have taken his boys fishing more. But they both grew up to be fishermen. They both became good husbands, fathers and Godly men. Their kids became fishermen, too. They had a dad that took them. A Papaw too. There was no doubt in his mind that his grandkids would also take their kids fishing. He smiled once more and was proud. He hoped that more people would discover the magic of fishing and pass it on.

With the gentle rocking of the boat his eyes got heavy. A nap came easy. It was much needed rest. The hospital visits and all the medicine had taken its toll. Late afternoon he awoke to the screeching sounds of an eagle flying in the sky above him. It was out fishing, too.

As he lay there watching the eagle, he wished he had more time left. He thought to himself that he would go back to Canada fishing for walleye and pike with his son and grandson. Travel with his other son and grandsons to the Northwoods for those good-eating yellow perch. Going back to catch a snook or grouper in Tampa Bay or speckled trout at Gulf Shores would also be on his list. A limit of crappie, some trout fishing or maybe catfishing would be good too. Grabbing a mess of suckers and frying them up on the riverbank one more time sure would really be fun. He even thought about going wade fishing in a creek or sitting on the bank of a farm pond. Alaska salmon and halibut fishing was on his bucket list. So was fishing for redfish. It had never happened, and now there was not enough time.

The sunset was beautiful in the western sky. He watched as bats began their dance in the approaching darkness. It was feeding time. He listened to the owls and the whippoorwills as they started their nightly chorus. The smell of newly mown hay and someone's campfire drifted through the air. He knew he should be heading home. His wife would be worried. In the gathering dusk, he wanted to fish just a little longer.

The doctor had told him the radiation and chemo was not working. This was his last time to fish. He was at peace with that because he knew where he was going. He had messed up his life at times. He had made mistakes. But he had got his life straightened out and was walking the path he should have been all along. He wished he had more time to tell his wife and family he loved them and make more memories with them. He wished he had more time to tell others that no matter what they have done wrong in their life, they can still go where he is going.

The boat roared to life and he headed for his favorite fishing spot near the ramp to make another cast or maybe two. In the half-light he cast toward the bank. The topwater bait gurgled across the surface. A really big bass slammed it and the fight was on. When the battle was finally over, and he lifted it out of the water it was bigger than the one earlier in the day. He removed the bait from its huge mouth, lowered it back into the water and in the dim light watched it swim away. He looked up into the night sky filled with millions of stars and with a tear in his eye and a smile on his face said, "Thank you!"

"Just one more cast," he told himself. The lure hits the water. A fish engulfs it. The battle begins and then suddenly stops. He's snagged. The line snaps. "That's okay," he says to himself and smiles again. Too dark now to re-rig. It's time to go home. It was his last cast.

Fall

"Delicious autumn! My very soul is wedded to it and if I were a bird I would fly about the earth seeking the successive autumns."

– **George Eliot**

THE OLD BARN

The forecast for opening day of the firearms deer season was for rain with a chance of thunderstorms. My son was out of town and my grandson was at college in Kansas. It wouldn't be the same without them so why not just stay home? Wait a minute, this is opening morning I'm talking about. A tradition, for goodness sakes. How many years in a row have I enjoyed this special day? I had to be out there even if I was going to be by myself. Even if it was raining.

The alarm jarred me from my sleep. I got the coffee pot going, brushed my teeth, did my duty, grabbed my hunting clothes and rifle, filled my thermos and was out the door. I could see stars in the night sky so maybe, just maybe, the weatherman was wrong. My truck came to a stop at the metal gate on the gravel road and I got out to open it. No rain! I drove on down the road, crossed the creek and pulled up to the old barn sitting majestically in the field.

My plan was to leave the truck there and hike across the field to a tree where my stand waited for me. I got out of the truck and the thunder rumbled and lightning cracked and lit up the dark sky. I was sure thankful I had gone to the bathroom before leaving home. My hair would have stood on end if I had any.

I quickly decided I did not want to walk across a field with it lightning, while carrying a rifle, and go sit in a metal treestand. Then the sprinkles started, the thunder and lightning continued, and I got back in my truck. As I sat there thinking about what to do, the sky lit up again and it seemed like heaven opened. I swear I heard the angel chorus singing hallelujah and trumpets bugling. There before me was the answer that would save this day. I would deer hunt from the old barn's hay loft. My son, grandson and granddaughter had all taken deer from the old barn before, and so had I.

I jumped out of the truck, grabbed all my hunting stuff and ran inside. Then I remembered I had a folding chair I used when hunting in blinds still in the truck and ran back out to get it. The rain was getting heavier, but the old barn would keep me dry. It was still dark so I was in no hurry to climb up in the barn loft. With wide eyes I looked around and my headlight assured me there were no wild animals in the barn ready to attack me. I also made a mental note not to step in all the groundhog holes in the dirt floor.

The old barn was built over 100 years ago by a gentleman by the name of Christopher Columbus Meadows. I remembered the old black and white picture the owner of this land had shown me of Christopher Columbus holding a horse by the reigns and standing next to the barn.

My headlight shines on the big stacked rocks and hand hewn beams that are the foundation on which the old barn has stood for all of those over 100 years. I look at the ax marks on the wood and I see in my mind's eye Christopher chopping and shaping the log to become this foundation. I imagine him in the wooden wagon, pulled by the horse in the picture, going down to the creek to find the flat rocks for the beam to set on.

I look around at all the weathered wood that covers the

old barn. There was no electricity in this valley when the barn was built and wouldn't be for another thirty years or more. So how did they get this wood to build it? How has the wood lasted this long? There is no paint or sealant of any kind on it. Where did they get the old rusted hinges and nails? I will never know the answers.

My mind travels back in time, and I see the horse in the picture standing in a stall. I see corn stalks stacked in another area. Maybe this was where they milked the old cow. Is that daylight coming through the cracks? It sounds like the storm has let up. I better get up in the loft.

I climb the stairs that are just as sturdy as they were when they were built but step carefully around rotted boards on the loft floor. I set up in the big opening where they once brought hay up from below to be stored in the barn loft. My chair is comfortable. I pour a cup of coffee and stretch out my legs. This is a great way to hunt deer even if it's not raining.

I look around the old loft still amazed how they built the old barn this big and how it has stood this long. The owner tells me it's home to barn swallows, field rats, mice, a pair of black vultures that come here to raise babies every year and the groundhogs who made all the holes which will probably be the biggest reason the barn comes down someday.

The rain stops. Through my binoculars I see a buck by himself. He has a weird set of antlers. On the left it is normal but only three points. On the right it is short with two points and ugly. He walks slowly across the field with his head down. I figure all the bucks have teased him about his weird rack and the females don't want anything to do with this ugly buck.

I think for a moment about putting him out of his misery and click off the safety. But then I think maybe next year when he grows back a new set of antlers, they will be big and handsome. Then the ladies will be attracted to him and the bucks that made fun of him will regret it when he kicks their butt. I click on my safety.

Rain starts again. He will be the only deer I see this day but that's okay. I don't know why we have to get older to realize that deer hunting is not just about getting a big buck you can put out on social media to brag about. Deer hunting is about sunrises and sunsets, the wildlife that go about their daily routines not knowing you are there. It's about all the memories you make with family and friends or alone in a barn.

This day will be added to my store house of memories. Before I get too old, and as long as it remains standing, I would like to have a few more days hunting deer from the old barn.

BOONE & BO
THE AUTUMN OF THEIR YEARS

Bo was a beagle and beagles are usually rabbit dogs, but he was all squirrel dog and cared nothing about rabbits. He loved to chase and tree squirrels around the farm. Creeks, barbed wire fences, briar thickets or even a brake-squealing car couldn't keep him from his mission. Treeing squirrels was his job, and he was good at it. He would do it all day long before finally coming back home to supper. When finished he'd curl up on the porch completely worn out from his day's adventures.

Bo was my Grandpa Boone's dog, and both were as independent as they could be. Boone was in his eighties and I was twelve. He enjoyed taking his grandson hunting and was teaching me to be a squirrel hunter, too, when the time was right.

Back then we were poor so it was important not to waste a bullet. It would take a while to save enough to get more. Getting a couple of squirrels was our supper. With Bo's help that was never a problem. I still remember how good those fried squirrels and Grandma's homemade biscuits and gravy were.

Boone had worked hard on the farm all his life and his

heart was beginning to wear out. He slowed down a bit, took his medicine, and kept hunting squirrels with Bo and me. Like Boone, Bo was in the autumn of his years. Bo's gray muzzle reminded me of Boone's gray beard.

For Bo, chasing and treeing was the game. It was fun to watch him go after a squirrel. When it would run up a tree, he would climb part way up it in his excitement to get at it. As soon as he gave up climbing, he would sit at the base of the tree barking until Boone got there and shot the squirrel. A dead squirrel was not important to him anymore. He would trail up to a freshly killed squirrel and then take off after another one.

Our last day was perfect squirrel hunting weather. A crisp, clear morning had dawned when we reached the back forty of Boone's farm. Early sun sparkled on the frosted grass as we left the old truck. The trees were bare of leaves now.

Bo saw the movement of a squirrel and he went to work. Boone took a position by an oak tree and watched. He smiled with pleasure as he listened to the sounds of Bo. He held his old .22 rifle, still in mint condition, in the crook of his arm.

Bo was out of sight, but his bark told us he was after his quarry. His voice muffled as he chased it across a gully and it ran up a tree, as we knew it would. Boone walked slowly to the tree and prepared for the shot. The squirrel came into view out on a limb high up in the tree. Boone sighted down the barrel, but it moved slightly as Boone fired. The squirrel fell to the ground and then ran into a thicket of wild blackberries. Boone muttered to himself.

Bo was after him, but like Boone, slower than before. His voice high and clear, he started after the squirrel at a walk. As we watched, Bo fell. Quickly scrambling to his feet, he yodeled as he entered the thicket. He gave voice for another fifty yards or so and then there was silence.

I looked at Boone. His face was gray, his breathing was

heavy and his old face seemed more wrinkled. "Sit down Boone," I said. "He found the squirrel. I'll go get them." But Boone just stood there and didn't say anything.

I walked through the thicket toward the place where I'd last heard Bo. I found him stretched out, mouth open, eyes glazed. There was no life left in him. A couple of feet beyond his muzzle, the squirrel twitched and was still. I left them both and returned to Boone. He was leaning against a tree with his head bowed.

"I knew it when he fell," Boone whispered. We walked back to the truck, thinking our own thoughts. Boone broke the silence. "I hope to go like Bo, doing something I really like to do."

"I'll come back later with a shovel," I said.

"Thanks," Boone replied, "I don't think I could do it. One more thing though, would you bury the squirrel in front of him?"

I nodded as a tear ran down my cheek.

We got back to the truck and Boone reached in and got out an oiled rag and carefully wiped his old rifle and cased it. He handed the gun to me and said, "I don't think I'm going to hunt anymore. I want you to have it." In just a few months Boone was gone too.

I hunted for many years with Boone's gun and took a lot of squirrels with it. But it just wasn't the same without Boone by my side and the sounds of Bo treeing a squirrel. Today, the rifle sits in the gun safe in my office. I am now in the autumn of my years.

My sons grew up hunting squirrels with that gun. I taught them like Boone taught me. My grandson, Hunter, got his first squirrel with it after his Dad had taught him. There was never another dog like Bo, though.

When I am gone Boone's gun will be passed down to one of them. They all know the story of Boone and Bo in the autumn of their years and the story will be passed down too.

CHEAP MEAT

I told my wife right after we got married that taking up deer hunting would be a good way to provide us with cheap meat. I explained to her that I had gone several times when I was younger, but there were very few deer around back then. Since that time the conservation department had grown the deer population, so my odds of getting us some of that cheap meat was a whole lot better.

My rationale to her was to justify the purchase of a new deer rifle I wanted since I didn't have one at the time. She thought it was a good idea and surprised me with a deer rifle for a birthday present so I could go out and get us some of that cheap meat.

Let me just say that if you are trying to reason with your wife about taking up deer hunting to obtain a supply of cheap meat, don't buy a rifle. Instead, fence in the backyard, buy a beef calf, feed it for a season or three, and then butcher it yourself.

To help you understand what I am trying to say, here are some of the things I have purchased for deer hunting in the over fifty years we have been married. Besides the rifle my wife bought me, I have bought at least twenty other guns, but of course not all were for deer. I also won two really nice deer rifles at Elk Foundation banquets. Most are now being used by my sons and grandkids. I also have to add in my Elk Foundation, Turkey Federation and Ducks Unlimited memberships. Then there's the cost to go to their banquets and all the money I spent at their auctions and raffles.

Then there is all the ammunition I have purchased. I explained to my wife that I was saving us a lot of money by buying one of our sons all the reloading equipment he needed to help keep us in bullets. Besides, he really enjoys doing that.

Now, to make sure the guns and ammunition were stored in a secure place where they couldn't be stolen, it was necessary to get a gun safe for me and a gun safe for our son who lives nearby. She was fine with that because it also gave her a safe place to store her valuables.

Over fifty years of buying deer and turkey tags for myself, as well as my kids and grandkids, has to be added in. It also gave me great joy to make sure I had them all outfitted in all the latest and greatest camo clothing, binoculars, range finders, gun cases, treestands, blinds and a multitude of other must-have equipment.

I almost forgot that we also got into bow hunting for deer, so right now the count is eight average price bows, two rather expensive bows and two crossbows. Then of course I have to add in arrows, broadheads, quivers, cases, targets, new bowstrings and again so many accessories I can't even count them all.

Oh my gosh, I almost forgot game cameras, scent attractants, cover up scents, knives and other processing equipment, including a smoker to smoke deer snack sticks and summer sausage, and a dehydrator for doing deer jerky. Most of the time rather than butchering the deer ourselves, to save a little money, we took them all to a friend of mine to have them processed. Oops, forgot all the coolers we had to have to haul the processed meat in.

Speaking of hauling, we had to have two ATVs with

gun racks, plus a trailer so we could bring the cheap meat out of the woods. Then there were all the pickup trucks we have had to have including the latest which is a ¾ ton diesel pickup to bring the deer home to put in the freezer. On top of that is the gas, diesel, oil, tires and other maintenance, plus I almost forgot insurance and taxes. I just remembered the taxidermy work for numerous deer, and that was definitely not cheap.

I know I have to be forgetting something, but you are starting to get the idea. And all of this has been just for deer hunting. I am not even including all the necessary items we had to have for turkey hunting, dove hunting and waterfowl hunting. Did I mention the waterfowl boat we had to have? Now, we are also talking about getting into predator hunting. I don't even want to think about all the fishing stuff. Johnny Morris loves me!

For the amount of money spent on just deer hunting, my wife and I could be living in a very nice house, maybe even had a second home in Florida, been on numerous vacations anywhere we wanted to go and have plenty of money in the bank. Well, I may be exaggerating a little but not much.

It's all my wife's fault. If she had not got me that first deer rifle as a birthday present, this may have never happened. We could have dined on the best prime beef steaks and gone to the finest restaurants. But the taste of venison cannot be duplicated.

Neither can the time I have spent afield away from the hustle and bustle of everyday life not to mention all the peace and quiet my wife enjoyed with me out of the house.

Then there is the friendship that has grown between hunting buddies. The bonding time with my kids and grandkids and all the memories we have made together while hunting deer. It is all priceless and may have never happened if my wife had not given me the birthday deer rifle.

I truly believe deer hunting and all the other outdoor

activities my kids and grandkids have grown up doing helped shape them into the fine people they are today. My wife and I consider all the money we spent on them as an investment in their lives, and it has certainly paid off.

We are all proud of having supplied the venison to our family. Venison's unique flavor is just one of the many reasons we drag ourselves out of bed before dawn, stumble through the dark in the cold and climb a tree.

What it cost to get that deer meat has nothing to do with it, really. You see, venison is priceless, but it sure as heck isn't cheap meat.

STUMP SITTING

The fall air is crisp as I start my journey up Dogwood Mountain. It's really a big hill but I named it that because I love the mountains. Here in the Ozarks area of southern Missouri we don't have mountains like out west, just big hills. The dogwood part of its name comes from the hundreds of trees with their showy white blossoms that bring beauty to these hills in the spring.

For a moment I listen to the soothing sounds of water as it tumbles down Dogwood Mountain Falls and then glance over my shoulder as the sun starts peeking over the hills behind me. The curtain is rising and I'm not in my seat.

My pace quickens as I head up the trail that follows the falls, then turn onto another trail that winds its way to the top of the top. My leg muscles burn as I climb over rocky areas in the trail, but I continue on.

Finally, I see it. To some people it may be just an old stump where someone cut down a tree a long time ago, but to me it's like an old friend waiting at the end of the trail. I hurriedly remove my backpack, take out my thermos, and pour a cup of coffee. It's stump-sitting time once again.

From my stump I see a thin haze over the stream that winds through Bull Creek Valley. There's a hint of smoke in the air from the cabins and homes that dot the valley.

Crows calling to each other on a distant ridge, and a fox squirrel scurrying through nearby treetops break the silence.

I know that somewhere below turkeys have flown down from their roost and are feeding in the fields. A doe and her yearling have probably joined the turkeys as a buck watches them from his hiding place. The kingfisher squawks as he flies through the mist over the creek. He's fussing at a heron that's fishing for breakfast or maybe a busy beaver.

The sun rises higher and the show begins. The gray of the morning is suddenly changed to a kaleidoscope of colors. My eyes feast upon the bronze of the oaks, yellow of the maples, red of the dogwoods, and green of the cedars and pines. The blue of the sky and white of the fluffy fall clouds add to the special touch of nature's painting.

It's too bad more folks don't take time for stump sitting. In today's hurried, pressured, fast-paced world, stump sitting can be an escape for a little while. Good stump sitting time only comes once a year - in autumn. With the perfect autumn day at hand, the stump sitter can fully enjoy this special time.

When stump sitting, all things in life begin to take their proper place. Big things become little, little things become big. Somehow, stump sitting helps you forget about work that needs to be done, bills that need to be paid and all manners of other worries. They are all washed away by the cleansing action of stump sitting.

The sun is high now and good stump sitting time is gone. I finish off the last of the coffee, put the lid on the thermos and put it away in my pack. I look up and say thank you. Then I put on my back pack, take a deep breath and start back down the trail. After just a few yards, I stop and look back at the stump. Maybe tomorrow will be good stump sitting time once again. But if not, there's always next year, and my old friend will be there waiting.

EVERYONE NEEDS A SPECIAL PLACE

I close the book I was reading, lean back and watch the autumn leaves flutter through the air before landing on the glassy surface of the creek.

My eyes pick out a single leaf still clinging to the tree above me. It moves with the gentle breeze until a gust of wind causes it to relax its grip and start its dance to the water. The water slightly ripples when it lands, and it just sits there for a moment as if resting. Then the current grabs it and it floats away. I watch as long as I can then wonder how far it will travel until it reaches its final resting place.

With the leaf gone, my eyes turn to the beauty of the trees reflected in the water. My ears listen to the soothing sounds of flowing water. My mind wanders back to all the memories that have been made at this special place on this special creek.

It's called Bull Creek. It starts as a gurgling spring and winds its way for many miles through the hills and valleys of the Missouri Ozarks. It meanders along under rusted bridges, past limestone bluffs, old cemeteries, open fields, and a cabin on the bluff above the creek.

Near the cabin, rushing water had carved out a deep hole perfect for fishing, swimming, and snorkeling. It is here I now sit, book in hand, staring at the water, and thinking about all the memories.

Here is where one grandson and a granddaughter caught their first fish. Now they're grown and it won't be long before they are taking their kids to catch their first fish.

Spring rains nearly always flood the banks. The awesome power is something to behold and fear. Spring also means dogwoods, redbuds and wildflowers reflecting in the blue water. I always listened for sounds of peeper frogs and kingfishers announcing that spring is here.

As early summer arrives it was time to take the annual first swim of the year in this cold, spring-fed creek. The grandkids tradition was to push their papaw in and then laugh as I came up screaming and gasping for air. They always thought I was kidding but I wasn't.

As summer continued, this special place played host to family, friends, and neighbors. Fishing continued, air mattresses dotted the water and lawn chairs lined the banks.

Saving tadpoles trapped in little pools of water and moving them safely to the creek was a favorite grandkid's activity. Catching crawdads was enjoyed by young and old alike. Those "rotten" grandkids would laugh again when Papaw would get pinched by an upset crawdad.

The clear waters of Bull Creek made snorkeling a popular thing to do for everyone who visited. The underwater world is fascinating! Bluegill would swim right up to your face or nibble at you as you floated along in the water. Bass and hog suckers didn't want anything to do with these homo sapiens that had invaded their home and would skitter along ahead. Sun fish usually guarded their nest or hid back under a rock ledge. A multitude of colorful baitfish would swim around in schools continually battling the swift water.

I remember the time I snorkeled under the water and took some real lobster claws and placed them where they stuck out under a rock ledge so they would look like the granddaddy of all crawdads lurking under a rock. I then watched as my neighbor, Bob, snorkeled closer and closer to where I had hidden them. I still laugh when I think about the look on his face when he came up out of the water.

If you were really lucky or unlucky depending on your fear of snakes, you might even get the opportunity to swim along with a 4-foot long water snake. No, it wasn't a fake snake, and no, I am not scared of snakes. At least as long as I knew they weren't poisonous.

I was a little nervous once though. I dangled a crappie jig in front of this same snake and he struck at it and caught the hook in his mouth. I didn't have much experience unhooking a writhing, very mad water snake, and was sure thankful the line broke before I had to figure it out.

When it was hot outside and there was no one else around, there was nothing cooler or more relaxing than heading to the creek, sitting a lawn chair in the water under the shade of a big old tree and reading a good book. I could usually get through a couple of chapters before the rippling water lulled me to sleep.

As summer gave way to fall, I still enjoyed taking a book to the creek. If grandkids came down, we fished or had rock skipping contests. When our Wisconsin grandkids came, they liked to find rocks with holes in them or "beautiful" rocks on the gravel bar.

This was the time of year when you might surprise a pair of wood ducks as they paddled along in the water, catch a glimpse of a whitetail deer or wild turkey at the waters edge, or even see a beaver busily working on his winter home.

Trips became infrequent when winter arrived. Sometimes I would wrap up warm and go there to see the frozen water along the banks. I was always hopeful I would see an eagle perched in a tree or flying overhead. If the day was not too cold, I would sit down, enjoy the peaceful serenity, and think about all the things that took place there.

As the grandkids got older they stopped coming. This

special place had lost its magic to them. They would rather go boating on big lakes or do other things. Now they're off to college.

Grandma and I got older, too, so we eventually sold the cabin on the bluff to a young couple with twin six-year old girls. Now they make their own memories. I still come back once in a while to enjoy this special place on the creek.

The leaves continue to fall and now cover the water like a multi-colored blanket. I stand, put my book under my arm, wipe the tears from my eyes and soak in the beauty one more time before turning to get in my truck and drive back home to Grandma.

Everyone needs a special place to go make memories.

DEER CAMP CHARACTERS

Every deer camp has its cast of camp characters. Individuals with their own special uniqueness that, when blended together like spices and seasonings in a recipe, make deer camp so special.

My deer camp has Dean. A bundle of energy and wise cracks who hunts deer and moose and elk but is afraid of a little mouse. His mouse phobia has brought great joy to the rest of us camp characters. We've never seen anyone get out of a sleeping bag as fast as when a stuffed mouse "accidentally" got in the sleeping bag with him. I will never forget how high he climbed and the look of horror on his face when he thought the noise in the old wood stove was a rat instead of the bird it turned out to be.

You could call Dean our camp coordinator. He makes sure the camp cabin is properly stocked and clean, collects the dues, buys groceries, and helps hang stands. His most notable contribution is the annual Saturday night "boil". It's a grand feast of shrimp, kielbasa, mushrooms, broccoli, potatoes and corn on the cob boiled together in a big pot and dumped on the table for hungry hunters. He always cooks too much, but taking home a bag full of "boil" is part of deer camp.

Dean is constant movement, washing dishes, emptying

trash, picking up the cabin, bringing firewood in for the night, setting the alarm clock, and asking everyone where they will hunt the next day. His energy is endless, and he is always the last one in bed. The rest of us wouldn't want him to know it, but we don't know what we would do without him.

He may put up a front of being a tough, fun-loving guy, serious about his deer hunting, but I know the real Dean. He's the guy who takes his young son, Conrad, on a youth turkey hunt and cries when he gets his first gobbler. He's the guy who brings Conrad to deer camp, puts him in the stand with him, shares the moments of the hunt with him, and looks forward to the time when, once again, he will wipe tears from his eyes when Conrad takes his first deer.

He's the guy who helped my son when he first came to camp and took time to guide him on a successful turkey hunt one spring. He is also the guy who caused the lump in my throat when he showed up unexpectedly at my mother's funeral. He will never know how much that meant to me.

Conrad is the youngest of our deer camp characters and, like his dad, he's a bundle of energy and constant movement. I love his imagination. Computers, television and video games keep a lot of kids from developing an imagination. What a shame! When I was a kid, my imagination took me to the mountains where I trapped beaver and muskrats and fought wild Indians and grizzly bears. I don't know where Conrad's imagination takes him, but it will help shape his life, along with Mom and Dad, and maybe some deer camp characters, into the man he will be.

Our deer camp characters even include a celebrity, although I don't think he really considers himself one. Jerry co-hosts an outdoor TV show, has outdoor columns in magazines and newspapers, is a member of a pro-hunting team, and has hunted and fished around the world with country music stars, NASCAR legends and even generals.

I'm sure if you asked him, he would say, "I'm just an old country boy who has been lucky enough to get to do some things I never dreamed would be possible." I think he would tell you being in deer camp with the rest of us deer camp characters, and his son, Flint, or daughter, Chase, is one of his favorite places to be.

Jerry's been a teacher and a mentor to my son and me. To all of us, he's just ole' Jerry. To Flint and Chase, he's the dad who took the time to take them hunting and pass on his love and knowledge of the sport to them as they will to their kids one-day.

Then there's John or "J.B.", as we call him. Deer camp wouldn't be the same without J.B., his Wisconsin accent and holey underwear. Through his wise cracks and jokes, he doesn't fool me. He has a heart of gold. There's nothing fake about J.B. He is who he is. He'll never change, and I'm glad because I wouldn't want him any other way.

Ed is somewhat of a newcomer to hunting, but is quickly gaining knowledge and experience. As a business executive, he is under a lot of pressure and stress, so he looks forward to deer camp with the rest of us characters. He enjoys his time in the woods not caring whether he gets a deer or not. Ed was with Dean when Conrad got his first turkey and he, too, shed a tear. He delights in the hunting success of Daron and Flint and Chase. I will never forget my son's face when Ed passed on to him one of his still very good and very expensive bows. Like some of the rest of the characters, Ed's an old softy too!

Mike is the newest member of deer camp and adds his own uniqueness to the mix. He's the consummate outdoorsman, serious about his hunting with the knowledge to back it up. Slow to smile, he was the object of probably one of deer camp's best practical jokes.

Mike had taken a nice buck and brought it into camp with the adhesive tag around its antlers. Where he is from that's the way they tagged them, but in our state they must be tagged around the leg. Dean told him the rule and that he better switch the tag to the leg or it could be illegal. This was next to impossible without tearing up the tag. Dean and I left to check our deer at the local fire station and set up a mock arrest of Mike for "mis-tagging" a deer. Although Dean and I weren't there, those that were said the conservation agent played it perfectly. They even took a picture of a very serious looking Mike posed with his illegal deer thinking he was about to lose his hunting license, rifle and deer because he tagged his deer wrong. I'm sure Mike will find a way to get even.

The last member of our camp characters is a very special young man, my son, Daron. I am so thankful that Jerry got me in as a member of deer camp many years ago. If he hadn't, I'm not sure Daron would ever have gotten to take as many deer as he's harvested over the years and especially wouldn't have had the chance to take some of the quality bucks that now hang on his wall. Unlike most of the others, he doesn't drink, chew, smoke or cuss, but he sure enjoys being around all the deer camp characters, and they all think the world of him. Deer camp has brought us closer together as father and son and created memories that will last a lifetime.

I forgot to mention one camp character and that's me. I'm the "old man" of deer camp, the one who cooks the annual opening morning breakfast and helps Dean with his "boil". I'm the one who really doesn't care if I take a big buck and who is content to harvest does to help fill our quota. Most years I tag out as early as possible so I can use my ATV to help others get their deer out of the woods or help with deer drives. I look forward to deer camp every year and, in spite of the practical jokes and name calling thrown my way sometimes, it is important to me to be with the rest of the deer camp characters. It's more special for me because there are fewer deer camps left for me than the others.

Deer camps are not just about filling your deer tags. They're about wood ducks whistling through the trees or the sound of geese high overhead. They're about a wild turkey, a coyote or a bobcat happening by your secret hiding place. They're about two fawns playing chase underneath your treestand, squirrels rustling in the leaves, birds flittering through the treetops, sunrises and sunsets. They're about sitting around the campfire or the old wood stove and telling stories and jokes. Most of all, they're about sharing this special moment in time with your fellow deer camp characters.

A Family Tradition

It's early September. It won't be long now until the leaves start changing and the temperatures cool. When they do, something stirs deep within the souls of our family. It is an uncontrollable desire that takes over our mind and body. It's time to go hunting again.

We love hunting. Not for the killing of the game but for all the special moments and memories that come when we are out there in squirrel woods, dove fields, duck blinds, turkey blinds and deer stands. It is a family tradition.

I have special memories of hunting with our son, Daron. First it was rabbits and squirrels followed by doves. Then it was deer hunting and his first buck and then another and then another along with a lot of does and a few turkeys for the freezer.

Our other son, Kelly, never really had the desire to hunt. He tried but didn't like it, and that's okay. That was his choice. Later though, when he had kids of his own and our grandsons, Ty and Sam, wanted to hunt they tried deer and squirrel hunting together. Maybe when they grow up and have kids of their own, Grandpa Kelly will be out hunting with them. If they don't continue hunting that's okay too. Again, it will be their choice.

Daron's first child was a boy and they named him

Hunter. A name that turned out to be appropriate because he loves hunting as much as his papaw and his dad. I think about him as a three-year old little boy dressed in camouflage going turkey hunting with his dad and me for the very first time. I still picture him and his dad walking down a lane hand in hand, dressed in their camo and in perfect step with one another.

He's been in deer stands, squirrel woods, on duck water and in turkey blinds ever since. He has grown to understand, like his papaw and dad, that being out there in God's great outdoors is what it's all about whether you harvest game or not.

Hunter is now a man and will be graduating from college soon. While there, he met a special lady. Molly and Hunter hunt ducks, turkey and deer together. Knowing my grandson, I would say it won't be long until they are hunting elk, pronghorn antelope, mule deer, moose and bear together. Molly fits right into this hunting family. We all know that marriage is coming, and we will all be happy about that because we all love Molly. They will continue the hunting tradition and pass it on to their kids.

My granddaughter, Anna, is a cute, little petite young lady who loves doing all the girly stuff and shopping with Mom and Grandma. She's been a gymnast, took dance classes, performed in school and church plays, has shown cattle, been a pig judge for 4-H and a cheerleader. Oh yeah, she also shoots AR's and pistols and reloads ammunition.

She took her first deer at ten-years old and several more since, plus turkey. Several facts and pictures she and her dad will be sure be sure to share with the right young man when he comes along. I think big brother will be a little protective of her, too, as will her papaw. We will all want to know if the guy hunts or is willing to learn. Even if he doesn't, I bet Grandpa Daron will have his grandkids out there hunting.

I'm not telling you all this just to brag, well maybe a

little, but I do believe the outdoors has the power to change kids' lives for the better. So do parents and grandparents. Hunter and Anna are blessed to have the mom and dad they have just as Ty and Sam are blessed with their mom and dad. One day, if they don't already, they will realize that and be grateful.

Not because I'm prejudice or proud or anything, but they're all good kids. They are who they are because of the sacrifices, discipline and love from their mom and dad, as well as two sets of grandparents who love them very much and have always been there for them.

Hunting has been an important part of our family. The venison, waterfowl and other game has nourished our bodies. Time together out hunting has nourished our souls. I hope they will all remember the memories and create their own.

I have no doubt that our grandkids will all pass their love of hunting on to their kids and their kids; proud grandpas will be out there with them too just like I was. The family tradition of hunting will continue. I will look down from heaven and smile.

TIME AROUND A CAMPFIRE IS SOMETHING SPECIAL

There's something really special about time spent around a campfire. Smoke drifts away or gets in your face. Wood crackles and pops. Flames dance and flicker. Sparks float hypnotically upward into a dark night sky.

A campfire makes you feel better. It warms you to the bone. Magically it takes away stress and pressure no matter where it's at. It could be deer or turkey camp, on the banks of a river or lake, on top of a mountain or down in a valley, during a camping trip or in your backyard. It really doesn't matter because they're all magical.

Around campfires there are no TV's or electronic gadgets like smart phones, unless you're lucky enough to get service. Otherwise, just turn them off. There are only friends and family, or the quiet and solitude of being alone.

Campfires are for cooking food, lighting the night and keeping warm. They are for sharing memories of other times and other places, talking about loved ones and old friends who are no longer here, the big one that got away or missing the buck of a lifetime. We turn our backsides to the warmth of its flames but still shiver as our eyes widen listening to someone tell ghost stories.

Campfires are where grandkids roast marshmallows and share time with their papaw. They are a place to watch the flames dance as the worry of the work week melts away. They are a place for fish fries, cookouts and fellowship.

It's easy to sit and watch the flames play for hours while someone tells stories or you listen to night sounds. Flames of a campfire are soothing and always changing. As a campfire dies down to coals, the night slowly takes over and you know it's time to crawl into your sleeping bag or your own bed until morning comes.

To have a good campfire, you first have to know how to build one. Start by making a foundation of tinder using an old bird's nest, dryer lint, pine needles and cones, or get some fire cubes you can buy in your local outdoor store.

On top of the tinder, crisscross small pieces of kindling, like small twigs or thin pieces of wood scraps, making sure there is plenty of room for air circulation.

Now light your tinder from below, not on top, to get both it and the kindling going.

Keep adding kindling until you start getting a bed of coals, and then gradually add bigger pieces of wood while you still leave room for air circulation. Now sit back and enjoy your time around the campfire that you built.

Heat from a campfire is also used to cook food. The warmth of the food feeds your body from the inside which is the only real way to keep your body temperature up.

Campfire cooking should be done over a fire that has hot coals rather than flames. Flames have less heat and more soot which blackens pots. Coals also have a more even heat so food is cooked well. Food cooked over a campfire just tastes better. It could be a shore lunch on some Canadian lake, grilled venison at deer camp or just hot dogs and smores.

One of my favorite times around a campfire is in winter before the sun starts the day. While my wife still sleeps, I quietly head outside to build a campfire in the backyard fire pit. It doesn't matter how cold it is, and if it's snowing that's all the better, and I still go. Flames reflecting off the snow are beautiful.

The best time is when the sky is still dark and millions of stars still fill the night sky. The wood sizzles and pops, the flames dance and the smell of wood smoke drifts through the air. It's a quiet time. Not many people are up early like me. I warm myself by the fire and sip my coffee.

I think of my wife, my kids, my grandkids, my friends and how I am truly grateful for them. I think of my God and how much He has blessed me. I look up and thank Him for my family and the great outdoors that He created for us to enjoy and take care of.

I thank Him for time in a treestand watching sunrises through the trees and waiting for a deer to come by my secret hiding place. I thank Him that I am still thrilled to find a deer antler or a mushroom. I thank Him that a turkey gobble still gets my heart beating faster. I thank Him for the sounds of loons and elk bugles. I thank Him for time on the water catching fish or just paddling. I thank Him for campsites and campfires.

My thoughts turn to all the outdoor memories I have made with my kids and grandkids. I sure hope there are many more to come before God calls me home. I stir the fire, watch the sparks and wipe away a tear. Smoke must have gotten in my eyes. Time around a campfire is something special.

THE OLD MAN IN THE MIRROR

He was up early getting ready to pick up his son to go deer hunting. He had brushed his teeth and was washing his face. He paused to look at himself in the mirror and saw an old man staring back at him.

Maybe it was because his 74th birthday was on Christmas, and it would be here in a few more weeks. He stared at the old man in the mirror and saw wrinkles carved by frowns and smiles through the years of his life. He looked at the bags under his eyes. He saw his skin sagging down on both sides of his chin and what looked like a turkey wattle hanging below. What little hair he saw was gray. The old man in the mirror was in the winter of his life.

He pulled into his son's driveway and smiled as he loaded his deer hunting gear in the truck. He was proud of the husband and father his son had become. He moved over to let him drive. His old eyes didn't see as well in the dark anymore. The interior light of the truck revealed specks of gray in his son's hair. It was hard for him to believe that it wouldn't be long until his son would be a grandpa for the first time. He was in the fall of his life.

Not much was said as the truck traveled down the road to their hunting place. The son glanced over at his dad. He realized that his dad was getting older. He wondered how many more deer and turkey hunting trips they would have together. Dad was still very active, and his health seemed good. But, at his age, you just never know.

As he drove, his mind wandered to times when he was younger and Dad took him rabbit hunting, squirrel hunting and dove hunting. He thought of frog gigging trips, fishing trips and especially sucker grabbin'. Camping and trout fishing were fun, too. Deer hunting and deer camp together back then was awesome, but he never said anything.

He thought to himself how he needed to thank him for the time they had spent together in the outdoors and all the outdoor things he had done with his son and daughter when they were in the spring of their lives. This would be a good time to tell him how important all that was to him and them. They drove on in silence.

The truck came to a stop and the old man got out to open the gate. The dark night sky was getting lighter. They had to hurry to get to their stands before the deer started moving. They wished each other good luck and started off in opposite directions. The son stopped, turned around and watched his dad walking away until he was gone into the dark.

The old man got to his stand and started the climb up. It wasn't as easy as it used to be. He settled into his stand, got everything ready and sat in silence waiting. He thought about the old man in the mirror that morning and wondered how many more times he would be able to do this thing he loved so much. Right now, he still had the strength, the will and the desire, but he knew at his age that could change at any time. He didn't want to think about that anymore.

The dark turned to light and the wildlife started their day. Birds sang their songs, crows talked to each other and squirrels sounded like deer as they rustled about in the woods. He watched deer traveling through the frosted field below but out of range.

As the morning wore on, his thoughts turned to all the memories he had from being in the outdoors with his kids, grandkids and friends in the summer and fall of his life. He even thought of a time when he was fishing and would look over to watch his wife reading a book. He wished there had been more time spent in the outdoors with his son and grandsons that lived in another state. Where had the time gone? It went so fast. He looked up to the sky and said thank you for blessing him and forgiving him.

In another stand, in another place, his son sat waiting. He, too, had seen and heard the wildlife. He, too, had seen deer out of range and even a few that he let have a heartbeat for another day. He, too, thought about outdoor memories with Dad, his wife and his kids as well as the memories he would make with his grandkids someday. The outdoor traditions he loves would be passed on. He, too, looked up and said thank you. He even thought about how he was in the fall of his life and winter was coming.

There were no deer to field dress and load that day. They talked some on the way home, but it was mostly a silent trip again. The old man was thinking to himself how he wished his dad would have spent time with him in the outdoors, but he didn't. He thought about how he never heard his dad tell him that he loved him. He had no good memories from the spring of his life.

This would have been a perfect time to talk to each other about all the things they had been thinking about. Why is it so hard for men to look each other in the eye and tell them how they feel? A day will come when they will wish they had.

They pull into the driveway. Hunting gear is unloaded. The old man says, "I love you, Bub!" The son says, "I love you too," then watches until his dad has driven out of sight. He goes into the house, kisses his wife and goes into the bathroom to wash his hands. He looks in the mirror and sees the gray in his hair. His thoughts from the day sweep over him. He thinks of his dad being in the winter of his life. "I will be right back," he tells his wife. "I need to go tell Dad something."

AFTER THE FALL – SAYING GOODBYE

It sure is getting foggy. I'm not sure I could even see a deer sneaking through the woods in this stuff. Oh well, I just love being out here sitting in my stand even if I don't see a deer. It's a great time to be alone with God and thank Him for the opportunity to be out here in His great outdoors.

I wonder how many sunrises I have seen coming through the trees while sitting in a tree stand? After over fifty years of deer hunting it has to be a lot. I have watched a lot of sunsets, too, while up in a tree, but sunrises are my favorite. There's just something special about being in the dark watching the sun gradually bring light to the forest.

Hearing the first bird songs of the day is music to my ears. I even love the smell of decaying leaves on the forest floor. The first movement I see is usually a squirrel gathering nuts for the long winter ahead. It's amazing how much a squirrel sounds like a deer walking through the woods. Then there were the times I have watched a fox, a bobcat or some other animal traveling through, and they had no idea I was even there. There was also the time an owl thought the fur trapper's hat I was wearing on a cold winter day was breakfast and, with claws raised, dove at my head.

It's funny how we deer hunters tend to give our stands names too. Over the years I have sit in stands with names like Northwood's, Papaw Bear, Dad and Me, 23, Pond, Kelly, Red Neck and even one called No Name. Just thinking of the names brings back a lot of memories.

Most of my years sitting in those tree stands have been by myself, but the absolute best times were the years I shared them with my grandson, Hunter, while my son hunted with my granddaughter, Anna. Hunter got old enough to hunt in his own tree stand, and I am now once again sitting alone in the deer woods. It won't be too many more years and he will be hunting with his son or daughter and continuing to pass on the tradition. Just thinking about the good times when it was just him and me brings tears to my eyes.

When you sit there waiting for a deer to come by your secret hiding place thinking of all these things, you see them in your mind. Speaking of tears, as I sit here this day, for some strange reason, I am seeing my wife crying. The fog is lifting enough that I can now also see my sons, daughters-in-law and grandkids crying. What's going on?

"Honey, I love you. Why are you crying," I say to my wife? "Can't you hear me? Hunter, I know you have always had a tender heart but what's the matter, Bub? Don't cry Sis, your papaw's here. Ty, Sam, come here and give your papaw our secret hand squeeze and let me wipe away the tears. Kids, I am right over here!"

Hey, I also see some of my cousins and friends from church. There's Pastor Scotty too! What are they all doing here? I try talking to them and they act like they can't hear me or see me. Why is this room filled with all these flowers and pictures of me with my wife, kids and grandkids, plus pictures of me with fish and deer?

I hear someone ask my son how it happened. How what happened? My son, Kelly, chokes back a tear as my son, Daron, puts his arm around him to comfort him and he says, "Dad was always telling us to wear our harness and attach our lifeline when we got into a tree stand. He was hunting out of a ladder stand and for some reason, I guess he thought he didn't need to do what he always told us to do. He even wrote articles and did radio shows telling other people how important it was to do it, but that day he didn't. A ratchet strap broke; the stand slipped, and he fell out."

I fell out of my tree stand? I'm dead! You've got to be kidding! I have hunted that stand for years. My harness and lifeline were in my truck. I guess like most hunters, I thought this could never happen to me. I made a bad decision.

I say I am sorry to my wife for the times I have hurt her, tell her I love her one more time and that the boys will watch over her, but she doesn't hear me. I want to hug and kiss her, but I can't.

I stand right in front of my sons and tell them how proud I am of them for being the good husbands and fathers they are, but they don't see or hear me. I reach out to touch each of my grandkids, tell them I love them, and I am sorry I won't be there to watch them grow up and have families of their own, but they don't hear or see me either. I pray they won't forget their papaw. I hope they tell their kids about the memories we made together.

I feel a hand gently on my shoulder and a voice say, "I know this is hard, Larry, but they will be all right. God will watch over all of them for you. It's time to go to a better place. There are other people waiting for you when we get there, and I bet you have a bunch of fishing, hunting, kids and grandkids stories to tell them."

We turn to go but I look back over my shoulder at my friends and family one last time, and I say goodbye.

WINTER

"How many lessons of faith and beauty we should lose, if there were no winter in our year!"
— **Thomas Wentworth Higginson**

DEER MEMORIES

Deer season is just about over for another year. Archery season is still going on, but for most hunters it is done until next season.

Since mid-September I have pursued whitetails. Most of the time during the archery season, I sat in my stand waiting for a deer to approach. Some afternoons twenty deer went by before darkness ended the hunting day. Most seldom got within bow range. I did miss one shot and passed up several others.

Almost every buck and doe I saw during rifle season was within range, and I did take a couple of does for the freezer, but the biggest buck was a small eight, and I decided to give him a heartbeat and hoped I would see him next year.

I was out hunting again during muzzleloader season and passed on shooting several smaller bucks. Where the big bucks were, I have no idea. Even if they had come by my secret hiding place, I never know if I will take the shot until the exact moment they are within range. Sometimes it's a pretty good feeling to watch one walk away that I could have shot.

Big bucks are not why I hunt deer anymore. That used to be important to me, but I would rather see my grandkids', or some other kids', faces when they get their first deer. I love deer hunting, but I'm worried about it. The deer hunting tradition seems to be overshadowed by monster bucks. The quality of the hunt, in most cases, is measured by how the rack scored.

Big buck pictures are all over social media, the internet and TV. A lot of hunters feel so much pressure to take a big buck, I don't know how they can truly enjoy the hunt. When hunters spend more time in pursuit of a trophy buck than they do home with their family, that's not good.

When I hear of a youngster passing on shooting a doe because he thinks he has to take an 8- or 10-point buck so Dad won't be disappointed, I am sad. What have we done?

This year has reminded me that there are no sure things in deer hunting. Just when you think you know a deer's habits, they change. That is why deer hunting is a challenge and exactly why I enjoy it so much.

While I sat in my stand this year, I sighted geese and ducks flying over. I saw spectacular sunsets and colorful sunrises. I heard owls and crows calling. I watched rabbits, squirrels and turkeys feeding. I admired male cardinals perched in nearby trees. I felt the chilling wind, rain and weather too warm for deer hunting. These are the things I wish more deer hunters would also notice, and I know some do. They make time out there all very rewarding whether you get a deer or not.

I think what I enjoy most now, in my older age, is the quiet time alone up in a tree. It gives me time to think about all the memories I have from so many years of deer hunting with kids, grandkids, friends and even by myself.

My son and I spent a lot of years hunting with friends at a traditional deer camp. He has several really nice bucks hanging on walls in his home that came from that time. I couldn't begin to count the number of does we harvested that filled our freezers. We laughed, we played practical jokes, we ate good, we helped each other. It was a wonderful time and we made lots of memories. Sometimes when I am sitting in a tree I think about those good times.

When grandkids came along and got old enough to hunt, things changed. We hunted different land, and now it was my son and his daughter in their stand and my grandson and me in another.

Those memories bring tears to my eyes. As I sit there in a stand once again by myself, I think of their happy faces when they both got their first deer. I think of my tiny granddaughter in her too-big hunting clothes but not afraid to shoot a rifle. I think of quiet conversations with my grandson as we waited and the pride I felt as he grew into a very good deer hunter.

My grandson now hunts with his future wife, and my granddaughter is too busy with college and other things. I'm betting if she marries someone that hunts, she will be out there again. I am also betting that all my great-grandkids will be deer hunters too.

Another memory I often think about is when my other son, who lives in Wisconsin and had never wanted to hunt, called me to tell me that he and one of his sons wanted to take up deer hunting. His mom and I packed the truck with rifles, hunting clothes and deer stands and headed north.

The morning of the hunt we couldn't get my grandson out of bed, so my son and I went out together. I went to one stand and he went to another. Not too long after, I heard a shot from his direction. I texted him, "Was that you that shot?" He texted back, "I got one, Dad." I couldn't get out of that treestand fast enough. I don't care how old your kids are, it's a very emotional moment. My son was thirty-nine at the time and 6 foot 4 inches tall. I couldn't hug him enough that day. I was so proud for him, and he was so proud that he had done it when nobody thought he could or would. That is definitely a memory I will always carry in my heart.

Deer hunting should be about being out in God's great

outdoors and truly enjoying this special time with special people. Like the Native American Indians, we should show respect and reverence to the animals we harvest and teach others to do the same. We should also pass on this grand tradition of deer hunting to future generations. While you are doing all that, instead of worrying about big bucks, you too will make deer memories.

SERENDIPITY

I sit on a river gravel bar letting the sun soak its warmth deep into my bones. It's only December but it's already been a long winter, and it felt good. Birds were singing. Like me, they were tired of the cold, too, and were celebrating with song. The sound of flowing water blended with their chorus.

As my mind wanders, I poke around in rocks of all sizes and shapes that surround me. How long had they been there? Where had they come from? How did the holes get in some of the rocks? What are the fossils in some of them?

Did you know rocks are like clouds? If you look really close you see things in them. This one looks like the state of Texas, this one like a heart. Here's one that looks like Dolly Parton. Sometimes your mind sees crazy things when you sit alone on a gravel bar on a mild winter's day.

I stack all the "holey" rocks I find in a pile. Some will be slipped on to a length of wire and hung in trees around the house to serve as weather rocks. When you want to know what the weather is you just look outside at the rocks. If they are wet it's raining, if they are white it's snowing, if they are moving it's very, very windy. Others will be used to make things like mobiles, refrigerator magnets, handles for drawers, necklaces, bracelets, and whatever else my mind comes up with.

The rocks are dull shades of black, gray, tan, brown and white. Some sparkle when the light hits them just right. Most are worn smooth from being tumbled through the water. The flat, smooth rocks are what I am looking for now. These are "skipping" rocks.

I stand, stretch, and position my feet just right, look out at the water and, with a sidearm motion, send the first "skipping" rock flying across the water. Six skips! Not bad, but I can do better. Four skips! That was a bad throw. My feet must have slipped. Eight skips! That's better. As I continue to try and beat my record, I think about how I taught my sons to skip rocks, and they now teach their sons to be champion rock skippers. All kids need to learn how to skip rocks.

I bend down to pick up another "skipping" rock, and I see it. The sun is shining on it like a beacon guiding me to it. I kneel down, lift it from the gravel and hold it gently in my hand like a precious jewel. "An arrowhead, I found an arrowhead!" I shout silently to myself.

As I turn it in my hand, I think about the hands of the Native American who made it. How old was he? How long ago did he make it? What tribe was he from? Where did they live in this valley? What was it like back then? Was this used to take a rabbit or deer to help feed his family? This was a special moment. This was serendipity.

According to Webster, serendipity means to find something you were not looking for. Now, I would have never known that if not for a lady I know that is a big fan of Webster and knows the meaning of words I have never even heard of. The moment she said the word and gave me its meaning, I knew I had experienced serendipity several times in my life. The arrowhead was just another time I found something I was not looking for - serendipity.

On another unusually mild winter day several years ago, I was riding my ATV down an old logging road when a squirrel ran across in front of me. I followed the path of the squirrel as he ran through the woods. My eyes stopped at the sight of something white sticking up through the brown leaves. I hit the brakes on the ATV and backed up. What is that? Probably a limb or just some kind of fungus I thought.

Normally I would have just ridden on, but this day I walked toward the white shape to find out what it was. My heart skipped a beat when I saw it was the shed antler of an 8-point buck. As I held it, enjoying its beauty, I thought about how unique the antlers of a whitetail buck really are. Like fingerprints, no two are alike.

Since then, I have learned where and how to look for shed antlers and have found many of them. However, I will always remember the first one and the day I found something I wasn't looking for – serendipity!

I also remember hiking in the woods one winter. Spring woods are filled with wildflowers and budding leaves. Fall woods offer a kaleidoscope of color, and summer woods are ticks and chiggers and snakes.

Winter woods are quieter with the faint musky smell of decaying leaves. Trees and bushes are bare, allowing you to see things that might have gone unnoticed any other time. You might see icicles hanging off a rock bluff sparkling like diamonds in the sunlight, a bleached-out turtle shell, unusually shaped trees, animal tracks or even the animal that made them.

On this particular day, I suddenly realized I was walking along an old road bed. Trees had grown up in its path but if you looked hard enough, you could still see where others had gone many years ago. As I walked, I wondered who had used this road. Was it loggers, lead miners, soldiers, or people who lived here?

I kept following the old road until it crossed a dry creek bed. There, lying half buried in the gravel, was the metal rim off a wagon wheel. Here, at this place, a long time ago, they tried to cross this creek and the wagon wheel must have broken. In my mind's eye, I could see it happening. The wood of the wheel had long returned to the earth, but the rusted metal rim remained to be found by me when I wasn't looking for it – serendipity!

One spring, I was hiking to a special little waterfall deep in the woods that I liked to go to. I had been there many times, but this time I went a little different way than normal. As I followed the sounds of the water, I came by a big flat rock and sitting upright on the rock, was a soda pop bottle that dated back to the 1940's.

Like me, someone else enjoyed coming to the little waterfall. The bottle they had been drinking from had remained where they had set it for over sixty years. I came along and found it even though I wasn't looking for it – serendipity!

I hope there are many more serendipities to come in my life and in yours. Those special places, special people, special times and special things that come along when you are not looking for them – serendipity!

MOON WALKING

As I opened the door, left the warmth of our cabin, and stepped into the night, the winter cold greeted me. I zipped my coat up even higher, pulled on my gloves and snugged the stocking cap down over my ears.

This is a special night that sometimes comes only once a year. A time when the full moon arrives and reflects off the ground covered with the white of a winter snow and magically transforms the darkness into light. This was a time for moon walking.

I stand there for a moment transfixed by the beauty and then start down the path that leads to the old gravel road below to begin my adventure. When I reach the road I stop, suddenly realizing that the only sounds I hear is the water dribbling from the frozen waterfall to my left, the creek gurgling over the riffles to my right, and my own footsteps as they crunched through the crusted snow.

This road was once traveled by civil war soldiers, and I wonder if they, too, walked down this road, in the snow, on a moonlit night. I imagine them cold, hungry, wishing they were home, hoping to see their family again. Did they question themselves why they were even here fighting this war that pitted brother against brother, father against son? I am thankful I live now instead of then.

The sound of the creek calls me and I walk down to the water's edge. The reflection of the moon makes the water sparkle like thousands of tiny diamonds. I know the fish are there somewhere. Probably back under a ledge or a root wad, resting in their watery world.

If I walk down to the creek during the day, I might see a pair of beautiful wood ducks paddling around in some quiet water, a long-legged heron fishing for his dinner, a kingfisher flying down the creek squawking as he goes, or if I'm lucky, a bald eagle perched in a big sycamore tree. Tonight they rest waiting for the morning sun.

My mind strays and I think for a moment of the warmth of the little wood stove back at the cabin, but the old road pulls me like a magnet down its well-lit path. I only intend to go a little further and then turn back, but I keep going past ice-covered bluffs, the summer swimming hole, open fields, the swinging bridge, and finally, I stop at the old cemetery.

The moon lights up the cold, gray headstones as they stand out against a snowy background. Most are weathered so much you can barely read the names and dates. Some are just flat rocks with a name and date scratched in them. These are the people like the Keetons and Cobbs who settled this valley. They're the men who went off to fight in wars and those who struggled to survive here. Many died way too young. It was a tough time back then. I've been to the cemetery many times, but on this quiet night it was different.

My thoughts of those who lived here long ago was broken by a barred owl asking "Who, who, who cooks for you"? He was quickly answered by another owl on up the valley.

Time to head back down the road toward the cabin. As I start back, I pause to take in the beauty of the hills that surround our valley. They stand out clearly against the moonlit sky. I think about the black bear in his den in the old tree, turkeys on their roost and squirrels in their nests. I'll bet the deer on this bright night are up and about like I am.

I'm tempted to go through the woods back to the cabin instead of following the road. This night is so bright even the normally deep dark woods are lit up. I reason that I could easily follow trails I've cut along the bluffs and by the waterfall all the way back. It's late though and I can now see the cabin lights shining like a beacon bringing me back to my wife who's reading a good book and probably wondering where I am.

The woods will wait for another special time when the full moon is bright, the snow is on the ground and everything is right to experience the magic of moon walking.

SITTING ON A BIG FLAT ROCK

It's a warm day. For February that is. I'm sitting on a big, flat rock in the middle of the woods. The sun soaks deep into my bones. Days like this don't come that often in winter.

I take my jacket off, roll it into a ball, and stretch out on the rock.

Except for the sound of a deer mouse rustling through the dry leaves enjoying the warmth, too, or the occasional chatter of a squirrel, it's quiet here.

My eyes get heavy but just as I start drifting off, an old dead tree comes crashing to the ground and startles me back to reality. What is that old saying? If a tree falls in the woods and nobody is there to hear it, does it make a sound? My heartbeat slows back down to normal, and I stretch back out on the rock again.

You know, I hadn't noticed so many dead trees out here before. The wood eating insects must have got to them. Then the woodpeckers got to the insects. Then the holes the woodpeckers made became home to other birds and flying squirrels. Someday, when no one's around to hear it, they too will fall. Then mice will build nests in them, snakes will hibernate, and they will be a great place for storing nuts. Eventually though, they will return to the ground from which they came. Strange what you think about when you're lying on a big, flat rock in the middle of the woods on a warm winter day.

The musty smell of decaying leaves reminds me of how amazing nature really is. In another month or so, tiny buds will start appearing. Soon after, green leaves will burst out and unfurl. These woods, which now seem dead, will come to life again.

I will notice the buckeye tree, the first tree to leaf out, as I scout for turkeys or begin looking for mushrooms. The oaks, maples, hickories, walnuts, sycamores, and all the others will follow. Serviceberries, with their dainty white flowers, will be the first to bloom, followed by the redbuds with their tiny purplish flowers. The white blossoms of the dogwood are not far behind. Their colors add beauty to the spring woods. It's so much different than it is right now when, except for the brown leaves and green of the pines and cedars, I feel like I'm watching an old black and white television.

In summer, the fully leafed trees add cooling shade to these woods as I come here for morning hikes. Summer also brings ticks, chiggers and snakes so I'm not here as often as I am in other seasons.

As summer ends and fall begins, the chlorophyll that makes leaves green begins to break down and the true colors of the leaves are revealed. These woods become a kaleidoscope of red, gold, orange, and yellow. Trees drop their nuts to the ground and deer, turkey, and squirrels enjoy the bounty. Once again, I will be hiking, scouting, and hunting. It's my favorite season here.

But then, those same leaves that burst forth in spring will wither and fall to decompose and give nourishment to the same tree that gave them life. Beneath them, too, lies the seed of a maple, an acorn or a hickory nut waiting to sprout into another tree.

Wow! I sure wax poetically when I'm lying on a big,

flat rock in the middle of the woods on a warm winter day. If a man talks to himself in the woods, and no one's around, does anybody hear him?

My eyes sure are getting heavy again.

WHAT NOT TO DO ON A COLD WINTER DAY

It was that dreaded time of year. No, I don't mean tax time! It was the time that comes every year when there are no hunting seasons open and spring fishing is still weeks away. Normally I would at least go for a hike or something, but the wind was blowing so hard the snow was falling horizontally, not vertically, and the roads were slick.

There were even small dog warnings out. What that meant was if you had a dog like a Toy Poodle or a Chihuahua there were precautions you needed to take. If they needed to go outside to do their duty, they should be securely fastened to a long leash or they could actually be blown away into the next county by the wind.

A little while ago I thought my crazy old neighbor lady was out in her house coat trying to fly a kite in the wind. The kite was actually her dog "Poo Poo" and I wasn't really sure it was actually my neighbor because her house coat had blown up over her head. It was not a pretty sight. I was getting ready to run out and help her but, thank God, her husband came to her and "Poo Poo's" rescue. Now, I just have to get that terrible vision out of my head.

With all the excitement over, I decided this would be a good day to start getting my fishing stuff ready for the days to come with promises of beautiful sunny skies and gentle breezes. I comforted myself with the words some wise person once said, "No winter will last forever, and no spring skips its turn."

I started by going to the garage and bringing in all my rods and reels. It took several trips. My wife doesn't believe me that there are male and female fishing rods and reels, and they multiply when stored in a dark garage.

After checking over the rods for bent guides, cracks and such, I removed all the line from the reels and especially all those with bird nests in them. I sure wish the birds wouldn't do that. Next, I cleaned them up, oiled them and tightened screws.

Then out came the spools of line I had bought on sale three years ago. I wanted to make sure I wouldn't lose a fish again like that huge bass that broke my line right at the boat last year.

Some of my older rods and reels needed to be replaced, but that's hard to do when you've had them so long. They seem like an old friend. Besides, with some of the prices they want for the latest and greatest rods and reels, I would have to take out a small loan to pay for them.

I lovingly carried all my rods and reels back to the garage and threw them in a corner. Now it was time for several trips bringing in all my tackle boxes. My wife wasn't falling for the same story I gave her about the rods and reels. She just shook her head and left the room.

It is absolutely amazing what you can find in your tackle boxes when you sit down to straighten them out and clean them up on a cold winter day. Like the crank bait I found in one of them that had no rear treble hooks. It was because I had to have them cut off after I buried it in my forehead when I hung it up in a tree and tried to jerk it loose. I sure got some funny looks while walking from the boat ramp to my truck, and even stranger looks as I walked into the hospital emergency room. You would think they had never seen anyone with a crank bait hanging down between their eyes.

As I continued to rummage through tackle boxes, I came across a beat up old jitterbug. It was the one I had used to damage my boat and the snake that was trying to get in the boat with me. One of the hooks was bent out straight. I guess that happened when I accidentally hooked the snake and tried to cast it as far as I could out in the lake.

In more tackle boxes I found dried up worms, a half-eaten peanut butter sandwich, a melted candy bar, old fishing line, rusted hooks, assorted sinkers everywhere, empty pork rind jars, used Band-Aids and dried blood in several places. I don't remember how I cut myself. Maybe it was that snake's fault.

My exceptional mind thought it would be a good idea to take everything out of each tackle box and lay it out on the floor. That way I could wash out all the tackle boxes in the bathtub, dry them out with my wife's bath towels and then put everything back in each box organized by type of bait. That way, I would also know what I needed to repair or replace.

While I was doing that, I came across my favorite spinnerbait. You know, when I get to heaven, I want to take that spinnerbait with me. I shall cast with care into weedy waters where big bass lurk and a monster bass shall rise and strike my favorite spinnerbait and tail dance on top of the water. I shall fight it until it is near the boat, until the beautiful, blonde, long-legged, buxom woman I am with shall shout with fright at the sight of the monster bass. Net! I shall shout. Net! And this woman shall reach so swiftly for the net that I shall not lose the fish. That's not the way it happened last summer when my wife tried to help me, and heaven should be different shouldn't it?

Speaking of my wife, one thing you should never do when cleaning and organizing tackle boxes is, first of all, never wash out your tackle boxes in your wife's bathtub that she likes to take bubble baths in. Secondly never lay out your lures and hooks on her carpeted floor. Do you have any idea how hard it is to get hooks out of carpet without having a lot of snags everywhere? That, along with lots of stains that won't come out, means I will be replacing carpet in a few weeks. I will also be cutting out all expenditures I had planned for fishing tackle and even some fishing trips.

The next time we have a cold winter day, and I am looking for something to do, I think I will read a book, watch a movie, or take a nap. Cleaning out tackle boxes can get too expensive.

MY HOW THINGS HAVE CHANGED

My early years were spent on Grandma and Grandpa's farm. If you needed to go to the bathroom you walked twenty yards down a path to a little building that was outside the house and had no deodorizer. Toilet paper was usually the pages of old Sears and Roebuck catalogs, and you always checked down the hole for snakes and spiders before sitting down to do your duty.

Kerosene lanterns or candles lit the night because there was no electricity. There was no TV or phones back then either. Water came from a bucket we carried from the spring which also served as a refrigerator. Hauling hay for the animals was done with a pitchfork and a horse-drawn wagon. We slopped hogs and butchered them ourselves and hung them in the smokehouse. Milking cows was done by hand with a bucket and a stool. We drank the milk and churned the cow cream into butter.

Chickens were raised for their meat and eggs. I can still remember Grandma ringing a chicken's neck and watching it flop around. I can still smell the aroma of wet feathers as they were dipped in a bucket of boiling water to help make the plucking of feathers a whole lot easier.

Grandma cooked on a wood-burning stove. Everything

we ate was grown or made on the farm. We hunted and fished, not for fun, but to survive. Even at a young age my little single shot .22 sometimes meant the difference between having a supper of squirrel or rabbit or going hungry. A mess of bluegill caught with my cane pole and a worm was a special treat. We picked wild fruits like blackberries and gooseberries and gathered nuts. There were no supermarkets or fancy restaurants in those days.

There was no depending on the government to take care of us back then. There were no food lines and handouts to those in need. We took care of ourselves and worked hard. We struggled but we were proud of who we were, what we had and what we accomplished. It helped mold me into the person I became.

As a kid, besides hunting and fishing and working around the farm, my time was spent exploring the fields and forests. I climbed trees and rested in the comforting arms of their limbs, carved my initials in them and daydreamed. I imagined Indians hiding behind them waiting to attack me, rode my imaginary horse through the fields and climbed the hills in search of adventure. I camped out under the stars on summer nights. I captured lightning bugs and put them in a Mason jar with holes in the lid. I can still see all that in my mind's eye and feel them in my heart. I am a writer today because of it.

As I got older Grandpa let me hunt turkeys and quail with his old shotgun. He even taught me how to use his old muzzleloader rifle so I could hunt what few deer were around back then. Grandpa surprised me one year with an old bait casting rod and reel he traded for with a neighbor. Along with it came a rusted metal tackle box with some funny looking lures, and I became a "real" fisherman. A love for God's great outdoors was planted deep in my soul.

A lot of years have passed since my days of childhood and, yes, things have changed. I know my kids and grandkids have a hard time believing the stories I tell them of growing up on the farm. They don't think anything about it when they flip a switch and a light comes on or turn a handle and water comes out. They sure don't think about it when they flush a toilet, but I do!

I sit at my desk writing this on a computer that corrects my spelling and grammar. It stores all the articles I write, helps me do research, sends and receives messages, and I could keep going on because the list is endless.

My thoughts are interrupted by the morning news on the TV in my office. I have it on, not to watch all the bad news, but to check the weather forecast for an upcoming hunting trip with my son. I grab the remote and click the off button. If I want to know the weather, I can find that out on my computer or on my smart phone without listening to negative news and commercials.

Out in my garage and barn is all the latest and greatest hunting, fishing and camping "stuff". We have a bass boat with the newest electronics that do everything but hook the fish. There's a duck boat, ATVs and a 4-wheel drive truck to haul it all. My grandpa just wouldn't believe how things have changed.

I sit back in my chair for a moment and see memories on every wall. Fish, ducks, deer and turkey fans from some of my outdoor adventures. Antique outdoor equipment is also scattered about the room. Grandpa's old, rusted muzzleloader sits in a corner and so does his old fishing rod and tackle box. His old shotgun is in the gun safe next to my single shot .22 rifle.

On all the walls are pictures of kids and grandkids. Most of them are of their first fish or deer and times spent together with them in the outdoors. Among all the pictures and directly in front of me, as I look up from writing, is an old picture of Grandpa and Grandma's farmhouse where I grew up and where I was born on a Christmas Day. There weren't many hospitals back then either.

Some folks might say grandma and I have spoiled our

kids and grandkids. We have helped make sure they had the latest in electronics, clothing and anything else they needed for today's world. We have helped with vehicles and helped with college. They have all the latest in outdoor gear. We don't call it spoiling though, we call it making investments in the lives of good kids. They work, they get good grades and they are not into the bad things a lot of kids are today. We tell them we wouldn't be doing what we do for them if they weren't good kids.

Most of our investments, though, can be seen in the pictures on the walls. In case you don't know it, kids spell love T-I-M-E, and we gave our kids and grandkids plenty of that and still try to. So do our sons and daughters-in-law with our grandkids. Time investment has been taking them on lots of outdoor adventures throughout their lives. I have no doubt grandkids will do the same with their kids and grandkids.

My grandpa invested in me too. He gave me as much time as he could while trying to survive on that old farm. Maybe our kids and grandkids will have fond memories of us just like I have fond memories of my grandma and grandpa from a time long ago when things were a whole lot different than they are today. My how things have changed, but time invested in kids is still the most important thing you can do to make a difference in their lives.

TRACKS IN THE SNOW

In the quietness of the early morning, he sat staring out the window at icicles hanging from the roof. The same white scene greeted his eyes as it had for several weeks now. He got up and went to the kitchen to pour another cup of coffee. The outside thermometer showed that the temperature was in the single digits again as it had been for many mornings lately. At least it wasn't windy and causing below zero wind chills.

He loves watching shows like "Alaska...The Last Frontier", "Mountain Men", "Life Below Zero" and others. But, this was southern Missouri for goodness sakes. What happened to global warming?

As he stood there looking out the kitchen window sipping his coffee and staring at the cold, he watched birds coming into the feeders. The woodpeckers pecked at the frozen suet cakes. That's no problem for a woodpecker. Other birds pecked around anywhere they could find a seed. They needed the food to warm their little bodies. Among the birds were more bluebirds than the man had ever seen at one time. Usually he didn't see them until spring when they were ready to start nesting.

Suddenly all the birds scattered as a red-tailed hawk dove into the snow trying to catch breakfast. He missed and flew away probably thinking that catching a mouse would be easier. A friend had recently sent him a picture of a woodpecker frozen to a tree and another picture of a bluebird a friend of his had found frozen but managed to nurse it back to life. Winter is hard on those that have to live out in it every day.

The birds soon returned, and he made a mental note to put more bird feed out. He went back to his office. Most days in the past few weeks had been cloudy, dreary and depressing. But, this day the sun was shining and the snow sparkled like millions of tiny diamonds were scattered on top of it. His smart phone made a turkey sound and he picked it up to see several pictures of some special kids from church playing in the snow with big smiles on their faces. He and his wife had gifted them with their grandkids' sleds several years ago but there had never been enough snow to get out and have fun on them. Along with the pictures was a text from their dad that said, "They love it!!!!" and the man smiled.

He and his wife had been watching out the windows lately at their little neighbor buddy, Hudson, out playing in the snow with Mom, Dad and friends. Hudson also had one of their grandkids' old sleds. He, too, was enjoying it, and so will his sister, Lilly, when she gets big enough. Adults were having as much fun as the kids. The man smiled again thinking about it.

He looked out the window once more. In past days it had looked cold, cloudy and uninviting. With the sun shining and after watching the birds and thinking about the kids having so much fun, the snow suddenly seemed beautiful and inviting to him. He took his final sip of coffee, got up from his chair and started putting lots of clothes on. He figured if the Kilcher family, from his favorite TV show, could do it, and if those kids could get out in this kind of weather and have so much fun, he could get out and enjoy it too. After going through a pandemic during this past year, nothing seemed that hard anymore anyway. He knew that if it was deer season this wouldn't stop him from being out there in a treestand. He had even gone crappie fishing in this kind of weather. Besides, he had read somewhere that getting outside is good for your body and your soul no matter what kind of weather.

A turkey sound went off again and he picked up his phone to read a text from a friend. Knowing that he loved watching Alaska TV shows so much, the friend had sent him a story about a lady in Alaska who went to the outhouse, and when she sat down on the hole, a bear bit her on the butt. When her husband heard the screams and came running, a very stinky black bear came out from under the outhouse and ran off into the woods. Her husband successfully treated her wounds, and they will now have quite a story to tell their kids and grandkids. She probably won't be showing her scars though.

Since the man didn't have an outhouse, and black bears should still be hibernating, he chuckled and finished putting his clothes on. After putting another log on the fire, he ventured out into this winter wonderland. The first thing he did was feed the birds and put out a little water for them since everything was frozen. He then started a fire in his fire pit so if he got too cold, he could warm himself. Then, he reached in his pocket for his smart phone, clicked on the camera and started walking through the snow.

He was amazed at all the tracks he saw. There were a lot of bird tracks around the feeders, as well as tracks and a body print of a hawk who missed. Rabbit tracks led into bushy tall grass and also under a storage building. Squirrel tracks could be seen in the snow clinging to sides of trees and then across the snow to another tree and then another. Near their tracks were holes where they were looking for acorns. The tiny tracks were probably field mice. Deer tracks were on the hill behind the house near where the garden is in the spring. Dog or coyote tracks were there also. Raccoon tracks were on the dirt road behind. Tracks of little kids and sled tracks were nearby.

As he walked down the plowed driveway to the front of his house, he noticed something strange in the front yard. There were places with tracks and some disturbed snow but no tracks leading to or from them. Was it a mouse or a mole? Did a red-tailed hawk finally get a meal or two? Maybe it was aliens! The mystery may never be known.

He kept walking around taking lots of beautiful pictures of the snow and the sun glittering off the icicles. He saw even more squirrel, rabbit and deer tracks. The snow tracks were proof to him just how many wildlife critters also call this place home. You just never know what you will discover when you get outdoors away from the television and other electronics that steal so much of our time every day. May some of the tracks you find in the snow be your own.

CHANGING LIVES

It was right before Christmas break, and I was asked to give a talk about the outdoors at one of our local high schools. While giving my talk I could tell a few of the kids really enjoyed hearing about all the opportunities there are to get out and enjoy our great outdoors. It was comforting to know they had parents and grandparents who loved the outdoors like I do.

Most of the other kids acted disinterested. They were on their smart phones or were thinking about what they were going to be doing after school, flirting with the girls or just wishing they were somewhere else. Anywhere else! Having to sit there and listen to this old man talk was boring.

During a question-and-answer period, it suddenly struck me that the majority of these kids knew very little about the outdoors and the intricate workings of mother nature. These kids, like so many others, have grown up in an age of single-parent families or a mom and dad who both work and don't have time or take time to get them involved in activities like fishing, hunting, camping, hiking and other outdoor activities.

They have grown up with their smart phone, computer, video games, television or the local mall as a babysitter. They have grown up not knowing what it is like to spend the day exploring our many streams, lakes, fields and forest. They have grown up in an electronic gadget, fast food, disposable, throwaway society with little respect toward our wildlife, our environment and some, even their own parents and grandparents.

It is sad to say but we would probably be shocked to know how many of these kids were growing up in a world where our government has enabled generations of their families through assistance programs. Their parents have no desire to even try to work, as long as the government takes care of them with food stamps and their kids with discounted or free school lunches and now even breakfast programs. The system is broken.

As I drove home from the school that day, I thought about how kids are constantly searching for things that give their lives direction and a sense of belonging. I thought about how that search has led to teenage suicides, drugs and alcoholism instead of fundamental values such as love of family, God, respect for living things and creation. Values so basic to lasting contentment.

Questions began coming to my mind. How had my kids and grandkids avoided the problems of other youth? How had they grasped the understanding of basic values that kept them out of trouble? How had they turned out to be such good kids?

The answer was time. Time my wife and I spent with our sons instead of depending on baby-sitters. Time we spent taking them to church and to school activities. Time spent giving them a basic interest and respect of the outdoors. Time we spent listening to them and encouraging them. We were there for them in the good times as well as the bad.

Because we did that, when our sons grew up, they both married wives whose parents had done the same thing for them. It was natural for them to raise their kids and our grandkids up the same way they were. I have no doubt our great grandkids will be brought up the same way.

I also have no doubt that those families that depend on the government to take care of them will go on for generations doing the same thing they have always done. Something must be done to change that. Politicians say they are doing these programs to help them. I think they are doing them to enable these people and gain their votes so they can continue to be career politicians.

Driving on, my thoughts turned back to the kids at the school. My hope was that something I had said would encourage maybe just one of them to get involved in the outdoors and, in return, it would make a difference in their lives. My greater hope was that somehow the parents of these kids would realize just how important these young lives really are and that it is never too late to change their lives and make a difference.

Pulling into my driveway and seeing the Christmas lights decorating our house, inside and out, my thoughts turned to this wonderful time of year. As I sat there in my truck, I wondered how many young lives could be changed if their parents gift to them this Christmas was the gift of time together in our great outdoors and teaching them what the "Reason for the Season" really means. As I thought about that I smiled to myself because that would truly change lives.

A Christmas Letter from Papaw

Dear Grandkids,
I know most of your communication in today's world is through social media, and you don't like to read something as long as this letter, but please do! One day all the gifts you get will be gone but my Christmas letter to you will last forever.

As you continue your journey on this earth, always remember to keep God first, family second and all other things third. Let that be your guide and you will have a good life.

You will make mistakes and you will have problems, but those things help develop your character. Having the morality to always to do what is right, not just what is convenient, is important.

Never get too old or too cool to give your parents and grandparents a hug and tell them you love them. If it were not for their sacrifices and guidance, none of you would be the fine young people you have turned out to be. Someday you will have your own family. Always hug them and always tell them you love them.

I hope as you get older you will continue to discover the many wonders of nature like you have through these first years of your life. God created an amazing place for us get out and enjoy. It is worth much more than wealth and all the problems wealth can cause. It is also a wonderful place to escape and get away from the pressures of this crazy world we live in.

My wish for each of you is that God's great outdoors, in all its wonder, will always be an inseparable part of who you are. I hope you will always be amazed when you see a big buck sneaking through the woods near your secret hiding place, an eagle flying in a bright blue sky.

May you always be thrilled with the beauty created by magnificent sunrises and sunsets, the tapestry of colors in a fall woods, a field of wildflowers in spring, beautiful sights from a mountain trail and camping out under a million stars that light the dark night sky.

My wish for you is that you never get tired of the sounds of geese as they head south for the winter, a turkey's gobble in the spring, the haunting sound of a loon or the majestic bugle of an elk, bird songs filling the air, ducks coming into your decoys and the sound of water as you quietly paddle a river or lake.

May the smell of decaying leaves in a deer woods and campfire smoke around a tent or in your own backyard always bring back memories of simpler times in special places. I hope that the tug of a fish on the end of your line will always thrill you more than anything you could ever buy in a store or online.

Hunter, I hope you always remember catching crawdads, your first turkey hunt with me and Dad, time with your papaw in a tree stand, how proud Dad and I were for you when you got your first deer and an unforgettable fishing trip with me and Dad to the land of Canada.

Anna, I hope you always remember you and your papaw riding the ATV and singing songs, your first turkey, your first deer and the day I handed you my camera. You loved taking pictures of wildflowers, butterflies and other neat stuff and still do.

Ty and Sam, I hope you always remember riding ATVs, fishing in the Northwoods and at the Missouri cabin. Grandma and I loved the trips we made to Wisconsin bringing you bows, BB guns, pellet guns, deer rifles, hunting clothes and fishing equipment. I sometimes wish you didn't live so far away so we could have done even more outdoor things.

Always remember all the outdoor memories we have made together and that you have made with Mom, Dad and each other. I hope your future spouse will love the outdoors, or learn to, and together you will teach your kids and grandkids and go make more memories.

I know you are all busy, and even though my buddy, Ty, calls me the "old man" I am still ready to go make a few more outdoor memories with my grandkids. Call me, text me, Instagram me or whatever you do. You could even write me a letter.

Don't ever forget that Grandma and I are always here for you when you need us. Love you all to God and back!

Papaw

AN OUTDOORS LEGACY

My wife, Maryann, and I are proud of our family. Pictures of our kids and grandkids are in every room in the house, on our smart phones and in our hearts. The great outdoors has been a part of all our lives. She enjoys hiking and boat riding but leaves all the other outdoor things to me, her sons she loves so much and her precious grandkids. She has never complained about time or money spent in doing all our outdoor activities and is always there to support us.

Both of us grew up poor, but thankfully, my time was spent hiking around the fields and forests on my grandpa and grandma's farm. Grandpa taught me how to trap rabbits and how to take a squirrel with a homemade sling shot. Grandma taught me where I could find wild berries, mushrooms and nuts. I helped feed the family at a very young age.

I waded streams and caught fish with a bamboo pole. Grandma liked the addition of fish to our meals. I saved my pennies and bought a used single shot .22 rifle and a box of shells. Suddenly I was bringing more squirrels and rabbits home. I even took an old piece of canvas I found in the barn and some sticks and made my own tent to camp out under the stars.

When our boys, Daron and Kelly, were growing up, my wife and I struggled to make ends meet. There was not enough money to do all the things we wanted to do for them. Within me, though, was a desire to share the outdoors that I had learned on the old farm. When they were young, we took them fishing and camping as much as we could. As they got older, I was able to take them to do some rabbit, squirrel and dove hunting. Deer and turkey hunting finally started for Daron when he was a young adult and married to his wife LaVay. She loves to hike and camp and just smiles when she looks at all the deer, turkey mounts and pictures of her kids all over her house.

Their son, Hunter, majored in Wildlife and Outdoor Enterprise Management in college. He grew up hunting and fishing, so it's the perfect line of work for him. At college he found Molly and all of us are sure glad he did. She fishes and hunts with him. We all know that someday they will be making the big announcement that they will be spending the rest of their lives together. I have no doubt they will raise their kids to love hunting and fishing like they do.

I hope Molly tells their kids about catching her first bass, shooting her first duck and getting her first deer with their dad. I hope Hunter tells them about him and their mom flying across the water in a bass boat, sitting in a duck blind together and how he felt when she got her first deer.

Hunter has lots of hunting and fishing stories to tell his kids. Sitting on his dad's lap catching his first fish. Laying on the ground poking at bugs when Dad shot a big gobbler. Numerous times in a treestand with his papaw. His first squirrel, first deer, first big buck. The time he took deer with his bow, rifle and muzzleloader all in one year. From fishing in the creek by the cabin to winning college fishing tournaments.

His sister, Anna, loves her "Bubby". He helped her get her first turkey. They are good friends and were college roommates along with Hunter's dog, Maverick, better known as "Bubba", and sometimes Anna's rescue dog "Max". When that special guy comes along for Anna, I hope he loves the outdoors too. If not, I feel sure Anna and his father-in-law will get him hooked on the great outdoors.

Anna, too, will have plenty of stories to tell her kids. The deer she took with Dad by her side. The turkey with her brother. Catching fish with her dad and brother. Camping, hiking and kayaking with Mom and Dad. Riding the ATV with her papaw and discovering that she was a really good outdoor photographer.

Kelly's wife, Lexi, grew up in Wisconsin's great outdoors camping, hiking, boating and fishing. They have raised our two teenage grandsons, Ty and Sam, on lakes, rivers, hiking trails and campgrounds all across America and taught them to love the outdoors just like the rest of us. Papaw got them all the deer hunting gear they needed and the family is taking up ice fishing too. They also have a special place not far from their farm on the Wolf River that they go to fish, enjoy water sports and spend time together with family and friends.

I am confident that someday they will raise their kids to experience the outdoors too. Like Hunter and Anna, they will also tell their kids about all the outdoor adventures they had growing up. Fishing from boats and kayaks with Mom and Dad. Shooting a bow, rifle and pellet gun, traveling to places they had never seen before.

The memories we have all made outdoors, in the woods and on the water, are many. Our grandkids are all good kids. I believe with all my heart that the outdoors they love helped mold them all into the fine young people they are. That and having a great mom and dad to support them and instill in them a faith in God. Two sets of grandparents were also there for all of them.

I also believe they will all teach their kids and their kids will teach their kids to also enjoy the outdoors. An outdoors legacy will be passed on and continued through future generations. I can't help but wonder how many other lives could be changed and how the world might be different if more families created an outdoors legacy.

"Creating memories is a priceless gift. Memories will last a lifetime; things only a short-period of time."
Alyice Edrich

ABOUT THE AUTHOR

Larry Whiteley has been an award-winning outdoor writer for over 50 years. His family outdoor communications company produces articles, radio shows and social media heard and read all over the world.

For 33 years Larry has been the host of the award winning, internationally syndicated Bass Pro Shops Outdoor World Radio. For 30 years he has produced the nationally syndicated Outdoor World print tips for newspapers across America. His voice is heard on the overhead intercom and message-on-hold phone systems at all Bass Pro Shops and Cabela's store.

Larry's "from the heart" writing style appears in numerous magazines, newspapers and blogs. "I have truly been blessed to be able to write and talk about God's great outdoors," he says. "The awards and honors I have received are appreciated but none would have been possible without the gifts He has given me to use. He deserves the glory, not me! My hope from this book is that someone might be inspired to get out and discover the outdoors by themselves, with family or with friends and while doing so find God who created it all."

He and his wife Maryann live in Springfield, MO with family nearby and family in Wisconsin and Kansas but always in their hearts.

Made in the USA
Monee, IL
10 July 2021